*Robert Musil and the Question of Science*

*Studies in German Literature, Linguistics, and Culture*

# Robert Musil and the Question of Science

## Ethics, Aesthetics, and the Problem of the Two Cultures

Tim Mehigan

CAMDEN HOUSE
Rochester, New York

First published 2020
by Camden House

Camden House is an imprint of Boydell & Brewer Inc.
668 Mt. Hope Avenue, Rochester, NY 14620, USA
www.camden-house.com
and of Boydell & Brewer Limited
PO Box 9, Woodbridge, Suffolk IP12 3DF, UK
www.boydellandbrewer.com

ISBN-13: 978-1-64014-066-0
ISBN-10: 1-64014-066-2

**Library of Congress Cataloging-in-Publication Data**

Names: Mehigan, Timothy J., author.
Title: Robert Musil and the question of science : ethics, aesthetics, and the problem of the two cultures / Tim Mehigan.
Description: Rochester, New York : Camden House, 2020. | Series: Studies in German literature linguistics and culture | Several of this volume's chapters were reworked from previously published materials in German. | Includes bibliographical references and index.
Identifiers: LCCN 2019051201 | ISBN 9781640140660 (hardback ; acid-free paper) | ISBN 1640140662 (hardback ; acid-free paper) | ISBN 9781787446755 (pdf)
Subjects: LCSH: Musil, Robert, 1880–1942—Criticism and interpretation. | Musil, Robert, 1880–1942—Knowledge—Philosophy. | Musil, Robert, 1880–1942—Knowledge—Science.
Classification: LCC PT2625.U8 Z79843 2020 | DDC 833/.912—dc23
LC record available at https://lccn.loc.gov/2019051201

This publication is printed on acid-free paper.

Printed in the United States of America.

# Contents

# Contents

# Preface

Robert Musil (1880–1942) is one of the most important writers of the twentieth century. He is considered one of the great modernist writers whose intellectuality and breadth of vision is approached by only two others: James Joyce and Marcel Proust. Musil's distinctiveness even in the august company of Joyce and Proust, however, is that he was qualified in two technical fields—engineering and philosophy—and drew questions from these fields into the heart of his literary project. As a result, Musil, alone among writers of his rank of any era, was able to produce a set of reflections of profundity and sagacity bearing on the capacity of science to influence the experience of ordinary living. It is to these reflections that the present volume draws attention and from which it takes its point of departure.

This volume has had a complex evolution. Several of its chapters were reworked from previously published materials in German. I would like to thank Micheil Paton for providing me with reliable translations of these materials. As I reworked these materials in English and as the concerns of this volume made their presence felt, a transformation of varying proportions occurred. Though some chapters still betray the purposes for which they were originally crafted, all now respond—I hope in valuable ways—to the present task, which is to assay "the question of science" in Musil from differing but related perspectives.

This volume owes its existence to the generosity of two outstanding funding bodies in Germany, the Alexander von Humboldt-Stiftung and the Deutscher Akademischer Austauschdienst. These two organizations have helped sustain a scholarly life in German letters over more than three decades. Thanks are also due to two Australian universities (Melbourne and Queensland) and one New Zealand university (Otago) which, through material support and time release, have played an enabling role in the research that appears here. I am also grateful for the advice and support of colleagues in the Institute for Advanced Studies in the Humanities at the University of Queensland and for comment provided by anonymous reviewers during the final stages of preparation of the manuscript. Finally, I thank the many colleagues, too numerous to mention, whose work has contributed—even more than I am able to acknowledge in this volume—to my own understanding of Musil's legacy. This was a legacy, as the present volume seeks to make clear, that sought to value art and science as equally rich and valid languages of the world.

*A note on the use of sources*:
This volume caters to the needs of non-German speakers as well as German scholars. I have furnished my own English translations of Musil's works in all cases, as well as rendering secondary sources in English wherever possible. Where a translation might be considered contentious or specific to my own purposes, I refer the reader to single words, phrases, or entire passages in the original. Musil's works are cited from the standard Rowohlt edition of his collected works in nine volumes, using the abbreviation *GW*, the volume and page number. Citation from Musil's diaries follows a similar convention by using the abbreviation *TB*, the volume and page number. Musil's titles are rendered at first use in German as well as English, thereafter in English translation only.

As indicated above, some of the work that appears in this volume first saw the light of day in academic journals or collections. I acknowledge with gratitude the following journals and publishers: Metzler-Verlag (for material pertaining to the introduction), *Deutsche Vierteljahrsschrift* (in relation to chapter 2: "Musil, Ernst Mach, and the Problem of Causality"), Transcript-Verlag (chapter 3: "Musil's Theory of Vision"), Rodopi (chapter 4: "The Essays"), Königshausen & Neumann (chapter 5: "Modernity's Crisis"), the journal *Wirkendes Wort* (chapter 6: "Musil's Concept of Value"), Reclam (chapter 7: "*Unions*: 'An episode of more than merely personal significance'"), Peter Lang (chapter 8: "The Problem of Trust in *The Man Without Qualities*"), and the journal *Neue Beiträge zur Germanistik* published by the Iudicium Verlag (chapter 9: "Musil's Correspondence Project: *The Man Without Qualities* and *The Blackbird*). Full citation of this material is also provided in the bibliography.

The author
August 2019

# Introduction: Musil's Intellectual Position

Robert Musil's great unfinished masterpiece *The Man Without Qualities* (*Der Mann ohne Eigenschaften*, 1930; 1933) is, among other things, a novel about the impact of science on ordinary human experience. As the novel suggests in its opening chapter, science no longer inhabits the fringes of society; it is not an embroidery around human lives, a helpful, though unnecessary, contrivance. On the contrary, human lives, according to the focalizing perspective established in the novel, can no longer be imagined without science. Science has become one of the key languages that script modern life for us. At the very least, Musil's novel serves as a reminder that a set of larger historical arrangements has now come together to put science at the very center of modern life and all that we take that life to amount to.

For this reason, *The Man Without Qualities*, as well as most of the works that precede it in Musil's literary development, can be read as a meditation on these historical arrangements. On the one hand, these arrangements brought science out of the science laboratory and the academy and into the public eye, seemingly for the first time. On the other hand, the science on view in the opening sequences of Musil's great novel—the statistical science of probability—could only achieve such visibility after a long gestation extending over many centuries. Musil signals awareness of this gestation at several points in his diaries. For example, surveying the question of the certainty of scientific knowledge in the late 1930s, Musil notes in diary volume 10 that the facts of the physical world can never be known in any final way. At best, knowledge of the physical world can be considered only probable. By contrast, as he also observes, Cartesian method rejects probability as an acceptable goal of scientific procedure and is focused instead on statements of "certainty" (*TB* 1:525). These observations about the changing goals to which scientific inquiry has been directed over time indicate that the inquiry into science that Musil convenes in the major novel, as well as in earlier works, is arrived at on the basis of a deep acquaintance with the factors that had revolutionized scientific practice by the end of the nineteenth century. As Karl Corino, Musil's biographer, reports, it was the prospect of taking part in experimental work at one of the centers of the new scientific learning that enticed Musil to Berlin in 1903 to undertake graduate study with the philosopher Carl Stumpf in the first place (Corino 2005, 219). As Corino notes, Musil had identified Stumpf as working in the "tradition of Helmholtz." Hermann von Helmholtz's sign theory is a notable example of the importance of the new mathematical reasoning in fashioning the results of science.

After a period of preparatory training in psychology and philosophy at the Friedrich Wilhelm University in Berlin, Musil progressed to his doctoral dissertation under Stumpf's supervision in the years 1906 to 1908, focusing attention on the controversial epistemology of the physicist and philosopher Ernst Mach (Musil 1980).[1] The protracted encounter with Mach, though it involved Musil in certain difficulties not confined to epistemological questions, was to precipitate a lifelong engagement with Mach's thinking as well as with the wider revision of knowledge practices involved in the general turn against Cartesian premises. Musil was especially fascinated by Mach's suggestion that a functionalism tied to developments in the logical language of mathematics could circumvent mind-body dualism and deliver a transparent set of variables by which to assess human behavior (cf. Cassirer 1969, Pieper 2002).

Musil's fascination for Mach, which dates from as early as 1902 (Corino 309), soon led him into conflict with his supervisor, Carl Stumpf, who was clearly no enthusiast for Mach's theories (Gies 2003, 12–15). Stumpf clearly expected of his charge an equal understanding of the weaknesses attaching to Mach's epistemological assumptions. As Corino suggests of a draft version of the thesis Musil submitted to Stumpf in mid-1907, Mach had been treated altogether too kindly (Corino 310). Musil's fascination for the new thinking, of which Mach's philosophy was but one example, thus nearly derailed the dissertation project altogether. When we review the final version of Musil's doctoral thesis today, we must understand it above all as a compromise between Musil's own sympathies toward Mach's psychophysical monism and his supervisor's skepticism about this same monism. We can further surmise that the thesis was only rescued after a decision Musil arrived at—doubtless with great reluctance and after much soul-searching—to fall in behind the intellectual line his supervisor urged upon him. What finally resulted comes across as somewhat of a halfway house—a piece of analysis that rejects certain aspects of Mach's project without subduing the interest in Mach outright. Nevertheless, despite the outwardly critical approach to Mach taken by Musil in his doctoral thesis, Mach's thinking was to retain its appeal for Musil throughout his literary career.[2]

[1] See chapter 2 for a detailed discussion of Musil's approach to Mach in his doctoral dissertation.

[2] Pieper (2002, 14–15) notes three general areas of significance for Musil arising from Mach's theories: first, Mach's monistically premised functionalism; second, Mach's focus on the individual's personal experience of the world that militates against the contrast between self and world; and, third, the dissolution of ego-based forms of identity, especially in relation to the Other Condition in book 2 of Musil's novel *The Man Without Qualities.*

It was not until five years after the completion of his PhD, however, that Musil resolved to give himself completely over to writing. It is not too far-fetched to argue that the seeds for his eventual decision to forego the relative security of academic life in favor of the uncertain vocation of the writer were sown in Musil's partisanship in his dissertation for Mach's monistic approach to knowledge. Thus, even if it is granted that imaginative writing, by its very nature, must concern itself with the metaphysical in point of principle, perhaps even dominantly, it is also the case that Musil's innate preferences as a writer were to foreground precisely the opposite tendency, the inclination to discuss the merits of an "antimetaphysical" trajectory and thought—a tendency, in other words, to use the new probabilistic science as a tool with which to probe and develop understanding of complex life. The attraction for Mach's thinking that could not be expunged from his doctoral dissertation, then, was one of the earliest signs of an enduring intellectual project—a desire to give the scientific impulse in modern life its proper due. In this regard, Musil notes how scientific language has reshaped the language of the world, and in doing so, inevitably, the way that world is experienced. Since those who experience the world differently are especially those who have differently composed it through their science, the novelist begins by privileging the life experience of scientists. These scientists, by and large, become Musil's protagonists. As the vast literary experiment these protagonists bring to light asks, how has the fundamentally recomposed world language into which science—and, let it be said: scientizing and pseudoscience—have all penetrated altered the relation between science and art, between knowing and feeling, reason and mysticism? How then, in view of this recomposed life experience, is one to live?

## The Two Cultures

From the very beginning, then, Musil's instinct was to bring two different areas of knowledge into dialog—the sensibility of the writer on one side, the ambition of scientific knowledge to reflect with ever greater precision and clarity the essential shape of ordinary life on the other. In Musil's first novel *The Confusions of Young Törless* (*Die Verwirrungen des Zöglings Törleß*, 1906), the author has the protagonist run aground on the question of the imaginary numbers in mathematics. In an encounter that anticipates the much later discussion in the diaries about the proper object of scientific inquiry, Törless discovers that no certainty can be retrieved from mathematics, the scientific language of physics. The encounter with the mathematics teacher in his boarding school, whom Törless seeks out in hopes of dispelling certain "confusions," thus proves to be unsatisfactory. Törless finds in the pursuit of mathematical understanding no antidote to the conspiratorial bullying of a fellow student, in which he participates but which he equally abhors. Despite the fact that the encounter with imaginary numbers cannot shed any conclusive light—either for the main character or for the reader—on the nature of the problems Törless confronts in the novel, Musil remained unwavering in his convictions about the value of science, deploying the perceptual inversions of Gestalt theory in pursuit of an answer to the problem of sexual attraction in the two stories of *Unions* (*Vereinigungen*, 1911), foregrounding the probabilistic approach to experience in the experimentally conceived drama *The Enthusiasts* (*Die Schwärmer*, 1921), and, in the unfinished novel, *The Man Without Qualities*, finally undertaking an ambitious reckoning with Mach's thinking, and, more broadly, with the many new intellectual currents swirling about in the period the novel examines—the period immediately before the outbreak of the First World War.

In view of the comprehensive nature of this reckoning, one way to understand the primary concerns of *The Man Without Qualities*—a way highlighted in this volume—is to read the novel as an early contribution to a debate, nascent in Musil's time, that was concerned to weigh up the relative virtues of art, and the discourses of art and the literary, on one side, and science, and the nature and procedure of the scientific disclosure of the world, on the other. This debate has since been glossed as the "two cultures" debate—a term introduced by the English academic C. P. Snow in a series of lectures at Cambridge University in 1959. The contours of this debate, however, had been long anticipated in the German cultural context. These contours are visible as early as the 1890s in Wilhelm Dilthey's ruminations on the question of the organization of knowledge in the German university of the late nineteenth century (cf. Gadamer 2010, 235–46). Even before this, they were implicit in the discussions that led to the founding of the new German university in Berlin in 1810 on a model of *Bildung* or self-formation (cf. Wellmon 2015), itself a unifying response to the polarization of moral orders and natural orders, of "Kunst und Leben" (art and life), in the German cultural imagination around the turn to the nineteenth century. Despite such distant provenance, the question of the relative status of the two dominant claims to represent knowledge of the world gained greater, and perhaps its first genuine, potency in the modern era only after challenges were issued from within the domain of science against empiricist assumptions about the true nature of scientific inquiry. When, by the later nineteenth century, the crisis of empiricism in the natural sciences had become a concern not just for scientists but also for intellectuals more generally, the full implications of what would later become the "two cultures" debate finally came into view.

In the vivid terms that characterize this debate, there are "two cultures" of knowledge; and this is not just because there are rival practices and views concerning the proper approach to obtaining knowledge. There are two cultures because the methods to obtain knowledge and understanding in these two domains are ***starkly dissimilar and thus fundamentally disjoined*** from each other. An important informant about the nature of the bifurcation in knowledge signaled by the term "the two cultures," the German philosopher Hans-Georg Gadamer, has summarized the separation in knowledge in the following way: "In light of this situation it has to be admitted that the type of knowledge pursued in the human sciences [*geisteswissenschaftliche Erkenntnis*] is not that of the inductive sciences but has a quite differently premised [*andersartige*] 'objectivity' and is acquired quite differently" (Gadamer 2010, 246).

In Musil's time, the impact of science on the settled domain of the rival culture, art, was first registered in a powerful way in the form of Hugo von Hofmannsthal's short epistolary essay of 1902, the *Letter of Lord Chandos*. In this "letter," a fictional author from the seventeenth century, the eponymous Lord Chandos, addresses himself to a friend and confidant, Francis Bacon, a real historical figure and one of the most significant progenitors of the (first) scientific revolution in England, in an attempt to explain the severely disabling case of writer's block from which he suffers. The *Letter* takes the form of an existential lament. The historical addressee, a partisan of a "great instauration" of new knowledge in the seventeenth century, is implicated in the causes of crisis, though no explicit charge to this effect is laid. Yet if we translate the letter to Hofmannsthal's present, we easily see that it is not the first scientific revolution of the seventeenth century that lies behind the fictional author's crisis, but the second scientific revolution—the turn against empiricism in science by scientists themselves. For it is under the terms of the second revolution in science that the bond between the work of the scientist and the work of the humanist and the writer is severed. If the work of the scientist is to be taken seriously and made generally available for the benefit of society, then it seems obvious that the mind of the scientist, neutralized against humanistic concepts such as empathy, compassion, and spirit in order to do its work, must also be made available to that society. If the artist proceeds to inhabit that mind in an attempt to comprehend its value, the full gamut of stabilizing concepts offered by an older order of humanism immediately must be seen to collapse. The collapse of the old humanism, then, is the crisis that is announced to the wider literary world in the *Chandos Letter*. In terms of cultural discussions today, the *Chandos Letter* presents one of the first expressions—and possibly the first programmatic expression—of a new order of "posthumanism" in modern literature (for further information on the impact of posthumanism on intellectual discussions in the modern era, see Landgraf et al. 2019).

Two things may immediately be said about this beckoning posthumanist world. On the one hand, as the *Chandos Letter* makes clear, there are no anchoring master concepts left to orient thinking and behavior. As Hermann Bahr put it in Musil's day, following the philosophical insights of Ernst Mach, the "I" or "self" had suddenly to be declared "unsalvageable" (cf. Bahr 1981). The second thing that may be noted is the fact that the world that is to replace the older humanism has yet to issue a set of instructions by which to organize its navigation. The posthumanist world, in other words, does not yet have any terms by which it could be rendered moral, ethical, or even properly aesthetic. This is the world Musil begins to lay out in the two novellas of *Unions* and, even more pithily and memorably, in the miniature piece "The Flypaper" (Das Fliegenpapier) of 1913. In these early anticipations of the type of art to which Musil would soon develop an overriding commitment, the attempt to give expression to the mind of the scientist increasingly takes center stage. In the later major novel, *The Man Without Qualities*, Musil turns his protagonist into a mathematician—and not just any mathematician, but one suffering an intellectual crisis precisely on account of the fact that this denizen of the new scientific world is obliged to accept that there can be no path back to the old humanistic viewpoint. This world, as the novel reveals in the last chapters of the first book, has, with the death of the protagonist's father, already faded away. Yet the novel is not concerned to plot a path back in time to something allegedly more reliable. Rather, it may be understood, in line with Musil's entire outlook as a writer, as the attempt to give expression through literature to the terms of the coming posthumanism—a posthumanism according to which the language of ordinary existence, and the conceptual assumptions informing that language, no longer suffice for the basic decipherment of that existence, let alone for the living of it.

From this viewpoint, Musil's intellectual enterprise takes up the challenge that Hofmannsthal issues in the *Chandos Letter*. Science may well be implicated in the crisis that has called forth a severe case of writer's block—code, of course, for a wider suffering of the mind and heart that leaves Chandos, and by implication the author Hofmannsthal himself, bereft of ways in which to reflect the changing nature of ordinary experience. On the other hand, the language of science offers a potential way forward, a way to liberate that mind and heart from its dependence on older concepts that no longer provide orientation, as well as a way to broker new concepts that will better inform the future business of living. Musil, of course, has no certainty that these new concepts—should they be located—will provide a more effective navigation of daily life than that which they have replaced. But a tribunal on their prospects is certainly called for, and it is this tribunal, whatever its outcome, that provides another way in which to understand the broader "question of science" to which Musil draws attention.

## Functionalism

Before Musil made the decision to commit himself to a career as a writer, he had completed an engineering degree that included a component focusing on mathematics. The most detailed assessment of the author's acuity in this regard has been provided by Kollmann. Although Kollmann (2007, 12) certainly concedes that Musil should not be accorded the status of mathematician, he nevertheless recognizes that Musil did in fact acquire significant knowledge both in the field of mathematics and in that of theoretical physics. Kollmann (15) further observes that a conversation between the protagonist Törless and his fellow student Beineberg in Musil's first novel may be taken as confirmation of Musil's understanding of the importance of the new mathematics at the beginning of the twentieth century.

Beyond this non-trivial contact with mathematics, Musil may well have consulted an additional informant about the significance of the changes that had revolutionized the practice of science by the second half of the nineteenth century. Sebastian (2005, 77) has pointed out that Musil and the philosopher Ernst Cassirer had more than a surface acquaintance at this time. On the basis of this contact, as well as Musil's independently obtained knowledge of mathematics, it seems reasonable to conclude that Musil shared Cassirer's insights about the eclipse of the notion of substance in physics and its replacement with that of function (cf. Cassirer 1969). It can further be surmised that Musil had become aware—no later than 1908 when he began work on the novella project *Unions*—of the emerging issue of the two cultures that this dis- and replacement logically gave rise to. Moreover, Musil's detailed diary entries on scientific and technical topics attest to the fact that he now saw it to be of pivotal importance to tax the formal resources of literature in order to give an extended account of the problem of the two cultures, whose genuine importance could no longer be denied. One of the notable features of *The Man Without Qualities*, for this reason, is its mix of essay and traditional narrative—the new vehicle by which reflection on scientific themes was to be advanced in the realm of literature.

Henri Arvon was among the first scholars of Musil's works to point to the importance of the functionalist viewpoint for the author. In his 1970 study Arvon understood Musil's interest in functionalism as a way of opposing positivism, a school of thought associated with the mid-nineteenth-century French thinker Auguste Comte. Positivism, for its part, aimed to resist the extravagant speculations of earlier metaphysics, which had failed to give proper weight to the testimony of the senses and, indeed, had been entirely uninterested in doing so. From this angle, Musil's partisanship for Mach's psycho-physicalism served two purposes. First, it afforded a critical view of Comte's positivism, an outgrowth of eighteenth-century empiricism. Secondly, it provided grounds for a general attack on the later logical positivism of the Vienna Circle of the 1920s, notwithstanding the fact that the Vienna Circle had seen in Mach a direct line of intellectual descent. Musil's entry into complex intellectual terrain was thus afforded by the rich legacy of Mach's monism. This legacy clearly cut two ways: toward functionalism and its general interest in empirical reality, on the one hand, and against strict empiricism, on account of the promise of probabilistic approaches to the real, on the other.

In the major novel, Musil is concerned to convene a tribunal on this legacy and the contradictions associated with it. His concern is thus, first and foremost, a critical one. It should not surprise us, for this reason, that Musil's deployment of conceits that stand in for the new thinking should introduce us to a good deal of complexity. Nor does Musil foreclose upon this complexity by attempting to resolve its contradictions in one direction or another. Under the heading of "Semblances Occur" in book 1 of *The Man Without Qualities*, for example, it is not possible to determine what the author finally decides about the intellectual contribution issuing from Mach's thought. Musil cultivates a studied irony about these matters and even extends on occasion into satire. But to see irony as critical distance, and therefore condemnation, would be misleading. Musil's tribunal about ideas eschews certainty about how ideas should ultimately be taken and whether, indeed, these ideas can be acted upon. The tribunal he convenes for this reason is less a court of critical opinion of the last resort than an ongoing disputation that serves the goal of shedding light on its objects from a range of different viewpoints. Thus the "anti-phenomenalistic" lens through which the objects of the world are viewed in line with assumptions about "semblances occurring" does not condemn the appearances, dislodged from their knowable source in a "real" reality, so much as invite consideration of the problem that confronts interpretation when the final reality of things can no longer be ascertained. Following this approach, the operative factor in life is no longer the principle of reasonable grounds, of sufficient reason in the vein of Leibniz, erected on a standpoint of tacit (though unargued) realism, but rather the principle of unreasonable grounds, which is to say, skepticism about what the category of the real actually amounts to. According to this principle, as Musil tells us in this section of the novel, everything which ". . . in our real life, and what I mean by that is our personal life and our public-historical life . . ., actually has no proper basis [*rechten Grund*]" (*GW* 1:134).

Annette Gies, who has investigated Musil's use of the functionalist approach in some detail, gives this question a different emphasis, in the sense that she identifies in the unfinished last section of *The Man Without Qualities* two distinct types of knowledge that can be seen to result from the deployment of the functionalist viewpoint. She calls these, respectively, a "functional and an emotional type of knowing" (Gies 2003, 10). Gies refers to them as "two different methods of knowledge that build on each other" (10). Her analysis, moving away from the substantive question of the intellectual merit of the one "culture" in relation to the other, introduces a new, and possibly more fruitful, question into the novel—the question of how we should imagine the transition from one method of knowledge to the other, without concluding whether the one method is superior to the other. With this problem of transition in mind, it can be asked whether Musil came to regard the discrepancy between the two cultures as of lesser significance than the question of how best to utilize the resources of technique and technology, on one side, and culture and art, on the other. What, in other words, should the proper relation be between a functionalist comprehension of the world, the legacy of Mach's psycho-physicalism, and emotional literacy in the world, which rises from the ongoing need to live life according to the physical and mental comportment that shapes the ordinary perspectives of human beings? Did Musil find a way of negotiating this gap between the two cultures, as Mark Freed has argued (Freed 2010), on the basis of inductive models of reasoning and connective, essayistic utopias? I will return to these questions in later chapters of this volume.

## Positivism

In the most recent research on Musil, one finds an increasing tendency to treat Musil's unfinished novel as a form of scientific research. That is, the novel is no longer exclusively read as a literary work, or as a system underpinned by rules and tropes reserved exclusively for the realm of aesthetic literature, but rather as a (literary) phenomenon with positivist claims to knowledge, or as a document from which positive scientific knowledge can (or could) be obtained. Some scholars believe themselves able to discern in Musil's novel a particular set of scientifically based theories; these theories, in turn, are usually elevated to become the key explanatory model for the novel. The novel has been interpreted along these lines by Gnam (2001) and Hoheisel (2009) through the lens of the theory of relativity; by Emter (1985) and Könneker (2001) with reference to quantum theory; by Kochs (1996), from the viewpoint of chaos theory; and, finally, by Kassung (2001) under the heading of thermodynamic theories that cast the events of the novel in terms of the principle of entropy.

These comparative studies, as elaborately conceptualized as they each may be, weaken the claim of literature to be literature—plot, characters, symbolism, paradox, and other important elements of literature are sharply downgraded. In addition, associative tools such as analogies, similes, and metaphors are conjoined semantically to scientific concepts, the effect of which overall is to subvert the literary in favor of the scientific viewpoint. Musil's use of functionalism is understood in these studies to carry the level of scientific realism required in each approach. Consequently, the question whether Musil intended to critique functionalism or not is mostly swept aside.

The most impressive studies on the question of the "two cultures" in the novel pursue a course that is not directed at the achievement of positive knowledge, but rather understands the question of science in Musil to be a question regarding the possibility of transposing scientific ambitions toward precision into literature in such a way that everyday life could be enhanced and experienced both more consciously and with greater freedom. Seen in this way, scientific discourses in the major novel are not meant scientifically at all, but reflect instead an underlying problem of proper ethical praxis. This is the perspective developed in part 2 of this volume.

For now, it will suffice to point out that certain recent approaches to the question of science in Musil, especially in the work of Sebastian (2005), Pelmter (2008), and Adams (2011), have achieved genuine advances in our understanding of the question of science in Musil. These are studies that approach Musil's general orientation towards science as an endeavor ultimately attuned to the question of how the goals of scientific precision can gain application to ordinary experience. For Sebastian, for example, there can be no question of a sustained coincidence of the "two cultures," but rather only of a momentary "intersection." Sebastian sees the failure of the Kantian category of apperceptive unity as the main crisis moment the major novel is concerned to narrate (Sebastian 2005, 83). This moment becomes key to the "cognitive dissonance" of modern life the novel means to highlight. Pelmter, by contrast, considers the novel's references to science as ultimately being about the value of experimentalism—that is, the question about whether scientific experiments and experimental literature have points in common. Pelmter, arguing that Musil endorsed Mach's understanding of experimentation (2008, 64–67), answers this question in the affirmative. In a similar vein, Adams suggests that Musil looked on the accomplishments of mathematics as applicable in principle to the realm of literature. This should not be taken to imply that one has to become a mathematician. Rather, he argues, one should cultivate the habits of "a mathematical person" (Adams 2011, 75), which is to say, a person who is able to abandon ossified ways of approaching the business of living and, following the example of hypothetical probabilism in the new mathematics, cultivate new and greater openness to the many unexplored possibilities offered in life (133).

## The Structure of This Volume

The chapters of the present inquiry proceed from premises that assume that questions about science, as canvassed in the foregoing, can be validly put. This inquiry, indeed, contends that science both as an intellectual praxis and as a social program demands a proper reflection in art. Musil's art is memorable because of the depth to which it drives that reflection. This is not the only reason to appreciate Musil's literary projects, but the consistency, range, and quality of that reflection is surely a strong reason for holding Musil in high esteem.

The study is organized under two topic headings—the notion of the two cultures as understood by the author Musil and discernible in his works (part 1), and problems in aesthetics and ethics that come into view in light of the emblematic status of the two cultures in Musil's work and thought (part 2).

The first chapter of part 1 establishes the centrality of the question of science in Musil's works. The competing knowledge claims of scientifically led seeing and intuitive frameworks of knowing familiar to art and literature are examined in detail in this chapter. Chapter 2 continues discussion of the question of science by considering Musil's reading of Ernst Mach in the dissertation he completed at the University of Berlin in 1908. In this chapter I argue that it was the issue of scientific causality that formed the focus of Musil's dissertation. Though Musil found fault with Mach's epistemology, it was precisely this epistemology that was to serve as a stimulus for the portrait of "Kakania" that he was later to draw in *The Man Without Qualities.* This stimulus accordingly remained productive for Musil throughout his career as a writer. The way science sets out its view of objects is taken further in chapter 3, which explores, in relation to the short prose piece "Triëder," the extent to which Musil's narrative style is informed by, if not science outright, then at the very least something like a scientific atmosphere and a scientific lens through which to view the world. Chapter 4 reveals the dominant aspects of Musil's thought in relation to his vast output of essays—a body of work that reveals the significance Musil accorded to the historical moment his time was passing through. As this chapter discusses, the conclusion to the intellectual work carried out in the essays was the major novel itself, a work that sought to join the analytical fruits of discursive writing—as "essayism"—with those of holistically derived dispositions of feeling-response typical of the presentiments of art.

The chapters of part 2, though they follow the parameters of the framing discussion of part 1, have a different focus. Chapter 5 opens up an aesthetic discussion in regard to the artwork by considering the crisis that overcomes art once the question of science moves to center stage in social arguments about "the right way to live." Chapter 6 deals with Musil's concept of value, which I argue has been much misunderstood in Musil scholarship. As I discuss in this chapter, Musil's concept of value is shaped by influences taken above all from the thought of Friedrich Nietzsche. Nietzsche's importance for Musil has a number of implications, including a refocusing of the discussion of the two cultures and clarity about his understanding of ethics. Chapter 7 focuses on the important early novella project *Unions*, published in 1911. As already alluded to, Musil sought to carry out a form of scientific experimentation in this work. The Gestalt-style hypothesis the two narratives in the collection aim to test is that a quantitative sum of emotions, under the pressure of outward circumstances, can crystallize an emotional quality that will then lead to a course of action. The "unions" that are meant to arise from this experiment consolidate the focus on the ethical substrate in Musil's works. In chapter 8, the question of Musil's relation to systems theory is taken up. The portrait Musil draws of Kakanian society in *The Man Without Qualities* draws on systems theoretical impulses and concepts such as (systems-based) trust. This chapter asks whether the trust invested in the social institutions and formations of modern societies is sufficient to sustain relations between human beings at an ethical level. Chapter 9 takes this ethical discussion further, adducing an ethical underpinning or framework based on observed moments of similarity and correspondence between the inner mental and the outer physical world. In the late story *The Blackbird* (*Die Amsel*, 1928), republished in the collection of short prose *Posthumous Papers of a Living Author* (*Nachlaß zu Lebzeiten*) in 1936, these moments of correspondence are given exemplary status at the level of the individual, where the goals of ethics are seen to coincide exactly with the goals of aesthetics and storytelling.

The volume concludes with a discussion of Musil's continuing relevance for today's societies in light of the "fourth" digital revolution in human knowledge systems. As I argue in this final chapter, Musil's thinking not only anticipated this fourth revolution; it also provided clues about how an individual life that also means to be an ethical life can confront the challenge of this revolution. The chapter ends with suggestions about how the weakening partnership between the two cultures in today's societies might be reformed in line with Musil's thinking.

# Part I.

# Musil and the Two Cultures

# 1: The Question of Science

According to the noted literary critic James Wood, the novelist is always working with at least three different languages. These, he explains, are, first, "the author's own language, style, perceptual equipment, and so on"; second, "the character's presumed language, style, perceptual equipment"; and, finally, a third species of language Wood calls "the language of the world" (Wood 2008, 28–29). This third species of language is of an inherited type—it is what is already "out there" when novelists make reference to the world. It might be conceived of as the material substrate of fiction, though the author's imagining of it does not make it objective. The language of the world is, rather, a palpable effect of life that insinuates itself into the author's creative imagination. This imagination then renders it usable for the purposes of literature.

For the Austrian writer Robert Musil, whose major works were published in the period 1906 to 1936, the language of the world plays a particularly important role. Musil's last and unfinished major work, *The Man Without Qualities*, for example, sets out on its long journey of more than a thousand pages published during the author's lifetime and a thousand more from Musil's posthumous papers (which are included in the standard Rowohlt edition of the novel) not just by talking about the weather, which on the day in question is favorable and fairly unremarkable, but also by adducing a rich array of meteorological data. Along with details relating to barometric lows and highs, the description includes references to seasonal temperature differences and monthly fluctuations of temperature, the rising and setting of sun and moon, and the humidity of the air. The opening paragraph concludes with the remark: "In a word, that describes the factual rather well, even if it was also somewhat old-fashioned: it was a fine August day of the year 1913" (*GW* 1:9). The point of this description is to invite the reader to consider two types of language that have become available when talking about the world: the language of ordinary experience, that is, that which is referenced in the phrase "it was a fine August day," on the one hand, and the specialist language of science, as instanced by the meteorological data amounting to the same thing, on the other. The claim of each type of language to represent a form of legitimate experience, and the relation that obtains between these two levels of experience, are accordingly highlighted.

At the end of the opening chapter a traffic accident is recounted. The narrative, panning like a camera sequence from detail to detail, records, among a collection of other impressions, the sudden movement of a sharply braking lorry: the lorry veers onto the sidewalk pavement and comes to a stop, its driver jumps out, people gather like bees around the entrance to a hive, a man lies on the ground in the middle of the swarm—perhaps he is already dead. A couple watching the scene moves forward. The man says to his female companion: "'These heavy lorries, as they are used here, have an overly long braking distance.'" These words break the tension for the woman: "It was enough for her that this horrible incident was placed into an ordered context and was turned into a technical problem that no longer directly concerned her" (*GW* 1:11). The man, reporting as if in response to "a lawful and ordered occurrence," goes on to detail quite improbable statistics about the number of people killed and maimed in traffic accidents in the United States. The chapter ends with an exchange about whether the victim of the traffic accident is alive or dead.

Both the beginning of the novel's opening chapter and its conclusion, then, serve the purpose of shedding light on the author's understanding of the language of the world. This language, as these passages make clear, has been shaped decisively by technical language—by what is referred to later in the novel as a "technical way of being" (Technik des Seins; *GW* 1:131). This technical way of being is in turn linked to a problem that soon moves to the forefront of the novel's concerns. I will call this problem the question of science. This question is initially posed in the following way in the opening chapter: how does the scientific interest, which purports to describe general aspects of the state of the world, stand beside or qualify individually sensed experience, which by its very nature is exceptional and non-repeating—what the female passer-by senses to be "something particular" (*GW* 1:11)? The woman's male companion provides the first response to this question. It turns out that the exceptional event—the life and death matter of the traffic accident—can be talked of in two ways from the viewpoint of science: first, in relation to the braking distance of a particular make of lorry; second, in view of a broader statistical relation, when the traffic accident that has just occurred on a fine August day in the city of Vienna is compared with statistics concerning traffic accidents in the United States. The first way specifies a traditional mode of science, that of empirical measurement. The second way specifies a type of approximate measurement that was beginning to dominate Musil's age: that of statistical probability. Both modes profess to bring order to what is otherwise taken to be a "horrible incident" at the level of individual or subjective reference. Order—a psychological measure—is restored when a random event on the plane of human experience can be turned into a "technical problem" that "no longer concerns" those who witness it. The particularity itself and the interest in the fate of the victim become, strictly speaking, a secondary concern—discussion about whether the victim is alive or dead is relegated to the status of a post-scriptum remark in the chapter.

This, then, is the immediate context in which the question of science receives attention in the novel. It is unfolded from there in a number of different directions, as I hope to outline in this initial adumbration of the question of science in Musil's works and demonstrate at greater length in the argument of this book. Although the question of science undoubtedly becomes one of the novel's chief concerns, it would be wrong to suggest that this work, or indeed any other work of the author, somehow migrates into the province of science or needs science to make it intelligible (though some commentators have maintained precisely this, as was discussed in the introduction to this volume). This remains the case despite the fact that Musil underwent formal training at university in mathematics and engineering, as well as philosophy and psychology, and invented a chromatometer, or color wheel, for use in testing color perception in psychological experiments. The color wheel, incidentally, still bears his name ("Variationskreisel nach Musil"). Musil, in other words, was undoubtedly capable of "doing science" and, in other circumstances, might, like his father, a professor of engineering, have decided to embark on a career as a scientist. While Musil makes liberal reference to science throughout his works, beginning with an allusion to imaginary numbers in his first novel *The Confusions of Young Törless*,[1] his literary production—with the possible exception of one early work[2]—does not become a substitute for science or perform a work of disclosure in the same manner as the hypothetico-deductive experiments typical of science. Nor does Musil finally affirm the thought that fiction can be empirically objective, as Musil's coeval Sigmund Freud appears to have believed about the psychoanalysis he helped invent. In fact, Musil was dismissive of the scientific pretentions of early psychoanalysis, and rejected the easy causality about the emotions that he felt it promoted. Musil instead seems to have taken the view that empirical causality is not a defining feature of the writer's craft, even as the writer might also discuss science as an important agency by which to comprehend the world.[3] The writer's craft, Musil appears to have thought, is, at most, a species of complexity research, whereas science—at least the science in which strict causality plays the main part—aims to reduce its objects to ever smaller units of analysis. The "scaling

[1] Harry Goldgar was the first to attach significance to the deployment of imaginary numbers in Musil's first novel. In a reading that seeks to determine Musil's position in relation to psychoanalysis, Goldgar views the numbers as "a controlling symbol of the unconscious." Accordingly, he does not see references to imaginary numbers in that novel as part of an emerging discourse of science in Musil's works (Goldgar 1965).

[2] That work is *Unions* (*Vereinigungen*); I discuss it further below.

[3] I say more about Musil's nuanced relation to the notion of causality in chapter 2.

up" tendencies of literature thus oppose by definition the "scaling down" procedures of analytical science.

We must be clear, then, that Musil's fiction, though it meditates on the relation between science and life, scientific method and artistic practice, does not finally fuse the two in any idealization about how art might finally collapse into science and, say, as a species of phenomenology, thereby become "objective." Indeed, one of Musil's most significant formative experiences before he became established as a writer was to witness how Edmund Husserl's putatively objective phenomenology fails in its ambition to objectify.[4] It is likely that Musil abandoned the career in psychological science that had beckoned him precisely on the grounds that psychology—the psychology he studied as a doctoral student under Carl Stumpf's supervision at the University of Berlin in the early years of the twentieth century—was not able, and perhaps would never be able, to close the gap that yawned as a result of the disparate practices of art and science, the two inherited, but polarized, received languages of the world. The question of science in Musil's oeuvre, accordingly, is confined to assessing how science might be conscripted in the service of a more general question, itself of more visible importance in the major novel, the question, namely, of "das rechte Leben" (*GW* 1, 255), of the "right way" to live.

## Scientific Premises in Musil's Oeuvre

It is certainly true that the question of "the right way to live" in the major novel, and the question of the two opposed languages of the world lying beneath it, can be said to constitute the culmination of Musil's entire literary enterprise. This enterprise can be reconstructed in its essential outline from an examination of the earlier fiction. I have already mentioned the reference to imaginary numbers in Musil's first work, *The Confusions of Young Törless.* The imaginary numbers take on significance for a young protagonist for whom the precision of mathematics offers some prospect of release from the confusions he experiences in the school convent where he boards. These confusions occur when an internee of the convent steals from a fellow student and retribution for the crime is demanded by some of his classmates. As this retribution moves through ever more troubling moments of physical and psychological abuse—abuse that young Törless, the novel's protagonist, witnesses and even participates in—the realm of

[4] Musil's deep study of Husserl around 1904–5 is evident from his diaries. As diary entries from notebook 24 relating to Husserl suggest and other comments at this time reveal, Musil had already classified thought (and thoughts) not as anything clear and distinct in the manner of Descartes, but as "accidental, a symbol, . . . but like . . . the final link [*Endglied*] of an inner development" (*TB* 1:118–25).

pure mathematics offers not just a sanctuary, a retreat from life's tribulations, but indeed a way of imagining a solution to them. But this prospect of a solution runs aground on the imaginary numbers, which are not the porthole through which the source of the infinite might be glimpsed, as Törless perhaps wishes it, but, as the mathematics teacher explains, little more than a "Denknotwendigkeit,"[5] a necessary occurrence in an abstract system whose formalism imposes a set of logical relations that do not accord with actuality.

Musil's age, nevertheless, appears to have viewed the science of mathematics as providing precisely the kind of insight into the link between logical relations and the order of the universe that Törless ruminates upon in the novel. This robust interest of Musil's age in mathematics, and the quantum sciences emerging from it at the time Musil was writing, might on its own have contributed to the enthusiastic reception the novel received after its publication. One might see this interest as involving the sort of problem that quantum physicists such as Heisenberg, Planck, and Bohr had perceived with increasing clarity in the first two decades of the twentieth century—the problem, namely, that, despite its growing refinement and sophistication, science was increasingly obliged to acknowledge its failure to reliably observe, much less measure, the smallest units of physical matter. The solution to the age-old debate between Democritus and Plato seemed to hang on this problem of observation. The debate, following Heisenberg (2006), can be put in the following terms: are the smallest units known to human understanding, as Democritus first suggested in Greek antiquity (but naturally could not yet demonstrate), minute particles of physical matter called "atoms," or as science had begun to surmise in Musil's day, even smaller units subsisting below the atomic level that resist scientific observation but might nevertheless one day be objectively demonstrated? Or could there exist even below this atomic and subatomic layer of matter, as Plato thought, something that cannot lay claim to physicality at all but is better imagined as an "idea," a kind of immaterial cosmic force one might just as well think of as divine in origin, a primal thought animating and preceding all being and yet also causally involved in generating it?

For the German philosopher Immanuel Kant, writing in the late eighteenth century, this ancient Greek debate remained undecided and indeed appeared undecidable. In the antinomies of pure reason in the *Critique of Pure Reason* (Kant 1990, 2:412–512), Kant discussed four sets of dialectical inferences about the nature of the world that bore on the problem Democritus and Plato had debated. In the second of these

[5] These are the teacher's words in full: "'You must come to an acceptance of the fact that such mathematical concepts are purely conventions necessary for mathematical thought'" (*GW* 6:77).

antinomies, Democritus's view that "nothing anywhere exists save the simple or what is composed of the simple" (Kant 1990, 2:427; A 434/B 462[6]) is set against the Platonic argument that "there nowhere exists in the world anything simple" (A 435/B 463). In the fourth and last of the antinomies the view that "there belongs to the world, either as its part or as its cause, a being that is absolutely necessary" (A 452/B 480), a position that is consistent with that defended by Plato, is contrasted with the counter position, consistent with the position put forward by Democritus, that "an absolutely necessary being nowhere exists in the world, nor does it exist outside the world as its cause" (A 453/B 481). Kant's purpose in contrasting these inferences about the world in his first great *Critique* is to see them as instances of dialectical opposition that cancel each other out. An impasse or irresolvable opposition accordingly results. Instead of attempting to adjudicate the two positions, Kant sees them in the Transcendental Dialectic as a feature of thought that arises as a matter of course when certain all-encompassing cosmological arguments about the world are put forward. When the mathematics teacher whom Törless consults in relation to such cosmological matters refers him to Kant's *Critique of Pure Reason*, the instruction he is meant to find in this work relates to this insight about the nature of dialectical argument. The teaching that Törless is meant to draw from Kant's seminal work of philosophy is that cosmological arguments relating to the fundamental arrangement of the world are directly opposed by counter-positions that effectively annul them. These arguments therefore do not allow objective statements about the inner constitution of the world to be sustained, but tell us more about how reasoning works and where the limits of reason reside. The imaginary numbers, then, are constructions of this sort—a limit position (or *Denknotwendigkeit*) that results from the application of an abstract system of thought to the world that does not extend as far as an objective demonstration of any physical feature of that world. Thus, while the limit position might be necessary for an argument to take place within the terms of thinking, the feature of the world it seeks to identify is not. From this basic distinction between thought and reality Kant derives a further distinction between "noumenon" (the unknowable thing-in-itself) and "phenomenon" (the appearances of the thing-in-itself)—a distinction that underpins his ambition to refer all thought to the need to generate the "critical" practice of limit assignation and observance designed to take stock of the conditions by which anything at all can be known.

Törless, we are told in the novel, reads the recommended work of Kant's but puts it aside, having comprehended nothing of Kant's overall argument. Kant suggests of the ancient debate between Democritus and

[6] "A" and "B" refer to the 1781 and the 1787 editions of the *Critique of Pure Reason* respectively.

Plato that we cannot finally tell which party is right. We must therefore accept that rational thought has in-built restrictions that might not allow us final insight into the nature of the physical world we inhabit. Kant's project is to recommend a modest program of acceptance of those limits that apply when we exercise the privilege of thought: we should neither allow ourselves the luxury of "the slumber of *fictitious* conviction" nor the equally dispiriting despair of "the euthanasia of pure reason" (A 407/B 434). If there is a solution to the antinomies of pure reason, it lies with the need to comprehend the limits of reason and work out what legitimate knowledge amounts to for human beings.

For Törless, however, this sober finding has no effect. On the one hand, we might determine that Törless is too undeveloped in his thinking to accept the Kantian argument about the limits of reason, and too caught up in the emotional drama of pursuit and persecution practiced against the petty thief Basini to understand, much less heed, any domain limitations thought is allegedly subject to. But there is a second reason for his failure to gain purchase on Kant's philosophy after the interview with the mathematics teacher, and this failure concerns the scientific moment Musil's age passed through and to which the novel draws attention as a language of the world. If we return to the debate between Democritus and Plato we can reference this moment as a baseline uncertainty about whether Kant's position represents the last word in responses to the dispute. For has not the advance of science, a century after Kant, introduced new uncertainty to the question of the cosmological argument about the simplest units of matter and whether they are ultimately discernible to scientific observation? Might not it be said, in view of the quantum position in relation to matter emerging in the very years Musil was writing *The Confusions of Young Törless*, that a solution to the problem of matter might now be in view? Might science, as it moved forward on the basis of its own statistical revolution in mathematics, soon reach a definitive position about the nature of matter, and, secondarily, the question of the presence or absence of a causal force or "absolutely necessary being" animating it? Might we not, in short, come to a finding in favor of Democritus or Plato and thus reach a final position in the ancient debate that allows us to move beyond the limit position that held for Kant and his age?

A reading of *The Confusions of Young Törless* suggests that this last set of questions pertaining to the Democritus-Plato debate is purposely left open in the novel. At a time when Max Planck was moving to his suggestion about "quantum packets" in nature and Planck and Ernst Mach were soon to engage in protracted debate about whether atoms are indeed merely *Denknotwendigkeiten*, which is to say, necessary, but technically "fictional," assumptions about how to think about physical matter (Mach's position), or, alternatively, as Planck thought, real, empirically observable features of the physical universe, Musil's novel reflects the state of uncertainty about the nature of matter that had gripped his age. The gesture in the novel that is decisive in this regard is the decision to suspend judgment on this question about scientific observation and, in consequence, not ratify the Kantian "solution" to the problem that refers the human being back to the nature of thought and the dialectical position it is obliged to draw under the terms of cosmological argument: Törless, as indicated, puts to one side Kant's major work of clarification in relation to this question. It is noteworthy that this gesture is opposed by the stance of the mathematics teacher, who, no doubt cognizant of the statistical revolution that advances in mathematics have brought about, nevertheless refers Törless to the Kantian argument presumed still to govern these matters. That the novel sustains interest in the potential of turn-of-the-century quantum science to revise the Kantian position while casting such potential against the backdrop of an established philosophical position presumed to delimit it may be cited as an intellectual achievement in its own right. It is from the power of an intellectual position that refuses to foreclose on this scientific question that the examination of Musil's interest in the problem of the two cultures undertaken in this volume draws its impetus and derives its justification.

## Gestalt Theory

Musil's second work, the two concentrated novellas of *Unions* (1911), abandons the ground of formal mathematics for that of perception theory—an area of psychology Musil investigated under Carl Stumpf's supervision at the University of Berlin. These novellas mark a new phase in Musil's intellectual development because of their interest in how a psychological-perceptual problem complex focused on the human emotions might be explored through the means of literature. As the novellas then propose, this is the problem of how a change in the *quality* of self-perception is registered as a change in the *quantity* of smaller modifications that instantiate a new Gestalt formation on the emotional plane and, as a result, precipitate a change—a decision to act—at the level of judgment. Yet the literary experiment of the *Unions* did not meet with success, either for the readership that had been generated as a result of enthusiasm for Musil's first novel, *Törless*, or for critics such as Alfred Kerr, who had welcomed that novel but remained tight-lipped about its successor.

*Unions* is a significant moment in the development of the writer, I believe, because it brings Musil to the recognition that the truth claims of literature—even so radically experimental a literature as *Unions* aims to be—cannot even remotely resemble the truth claims of science. The arduous two-and-a-half-years of labor on this work, whose goal was the isolation of the inner quality of human emotion, then, ends the presentiment that literature might operate in the manner of science or function as a branch of it. This failure had important consequences for Musil. The intervention of the Great War initially brought a postponement of these consequences, but when Musil resumed the literary life after its conclusion in 1918 he resolved to move away from the *Unions* project and its more arcane methodological assumptions.

The short stories published under the title of *Three Women* (*Drei Frauen*) in 1924, as a result, have a conventional appearance; in fact, despite the use of historical settings in two instances, they now come across as classical expressions of literary modernism, of the modernist's engagement with the "new," particularly the new psychology with which Musil had had such special acquaintance. Yet even though these stories are more conventional in appearance, a strong interest in science can still be found in them. This interest is no longer directed at the question of whether the experiments of literature can resemble those of science (*Unions*), or whether science can solve any of the problems encountered in ordinary experience (*Törless*). The question posed in the third and some ways most valuable of these short stories, entitled "Tonka," is whether the scientific experience of life and the ordinary experience of it can be made to stand on the same ground. These questions become urgent at a moment when the scientist's girlfriend Tonka falls pregnant. A simple girl from the country animated by a poetic sense of life, Tonka denies having cheated on the scientist, and yet the scientist's reckoning indicates that a sexual encounter with another man must really have occurred while he was away on travels. At any rate he could not possibly be the child's father. A stalemate results in which the protagonist's scientific sense is challenged by another kind of calculation about life, at once "unprovable" and therefore materially unsound, but superior to material truth at the level of belief (it is this appeal to non-material truth that gives all three stories in the collection a neo-Romantic flavor). In attempting to construct a basis of trust on which his relationship with Tonka might be rescued, the protagonist would be called upon to do nothing less than abandon the entire project of Western scientific rationality—a project in which, as a scientist, he has come to vest great confidence. But the story is not resolved on grounds of trust or belief: Tonka dies of an infection contracted during her pregnancy—an infection that the calculus of probability indicates must have come from the protagonist himself. No overcoming of the separate domains of experience encompassed by poetry and by science is therefore countenanced.

Nevertheless, in the story "Tonka" Musil works out a problem complex that was never to leave him. This is the problem of how the ascent of science to a position of prominence in the social world has consequences for those who actually engage with it: science changes the way life and the living of it is actually experienced. This interest in the life experience of the scientist steps to the fore in *The Man Without Qualities*, where the protagonist, who might be called an elder brother of the protagonist of "Tonka," is by professional calling a mathematician. But it is also found in other works, such as the highly unconventional drama *The Enthusiasts* (1921), where science directly shapes the professional experience of several of the main characters, including that of the private detective who is asked to investigate a case of marital infidelity, and some of the miniature stories of the collection *Posthumous Papers of a Living Author* (1936), in which scientific perspective, which seeks to command an impartial ground of observation, fuses with the narrative perspective, most memorably, perhaps, in the miniature piece "The Flypaper." Here Musil experiments with a rigidly dispassionate literary style through the means of a narrator who observes the fate of flies landing on flypaper in strict adherence to an attitude of scientific neutrality. This story, one of the few written before the First World War in the collection, may be interpreted as a response to the new literary charter set out by Musil's compatriot Hugo von Hofmannsthal in his *Letter of Lord Chandos*. In this *Letter*, published in 1902, Hofmannsthal effectively called for a literature that would abandon the high-minded moral projects of the nineteenth century and instead follow and be inspired by the procedural example of empirical science (the letter is addressed to Francis Bacon, one of the main architects of the social aspirations of the Scientific Revolution in the seventeenth century). In "The Flypaper," it is the teeming life of the small world that accordingly absorbs attention, and, by extension, the habits of the human swarm that the narrator means finally to lay bare.

Notwithstanding these intentions, however, Hofmannsthal's *Chandos Letter*, which stages a crisis about the received language of the world, was not able to offer any blueprint for achieving a new language or a new literature. Despite the contention that art might be fashioned into a kind of second-order empirical science, as Hofmannsthal might be taken to imply, Musil's mature position from about 1919 or 1920 on is that art should be understood in its own terms as a special kind of undertaking—one shaped not by science but by the twin factors of aesthetic craftsmanship and ethical responsibility (Musil came to see these two factors, the search for style and the ethical undertaking, as being in essence the same thing). And while art might pursue scenarios of an experimental and seemingly mimetic kind, the experiments of art have no scientific status—they do not serve conclusively to prove or, to speak with Karl Popper, "falsify" any general or "real" hypothesis about the world. Art, in important ways, therefore, comes to occupy a different ground from that of science. In a landmark essay from 1918–19 entitled "Sketch of the Knowledge of the Writer" ("Skizze der Erkenntnis des Dichters") in which the emerging gap between what would only much later be called the "two cultures" is thematized, Musil assigns art to a realm of "analogical" understanding and practice, where a richly connotative, allusively symbolical, and perhaps only sometimes rational reasoning applies, and a narrowly logical, formally rationalistic, nature-directed, empirical reasoning of a type described in the essay as "ratioïd" is strictly avoided. Since, as the essay contends, there can be no question of art's merging with science on procedural grounds, the question of science serves a different purpose in Musil's oeuvre in general and in the major novel in particular.[7] This purpose concerns the manner in which science has reshaped the language of the world and the way that world is experienced. The question of science that is put to consideration in this volume, for this reason, involves a basic problem of separation between the experience of the exercise of scientific thinking, notably by scientists themselves, and experience of ordinary life that takes place at a remove from the scientific.

## The Question of Science in *The Man Without Qualities*

Nowhere are the dimensions of this problem of separation more evident than in chapter 116 of book 1 of the major novel.[8] This chapter is entitled "The Two Trees of Life and a Proposal to Establish a General Secretariat of Precision and Soul" (*GW* 2:583–600). The chapter is notable in part for the exchanges that take place between the novel's protagonist, Ulrich, and his chief adversary in the novel, a German industrialist by the name of Arnheim (who Musil conceded elsewhere was modelled on the industrialist and "world citizen" Walter Rathenau). Ulrich and Arnheim are adversaries in the sense that they have different conceptions about a cause that otherwise unites them: a cultural celebration linking the Austrian monarch's seventy-year reign in 1918 to the thirty-five-year reign of the German monarch due to take place in the same year (a time five years into the future in terms of the novel's temporal setting). Although hardly historical in any true sense of the word, the novel does make a notable use of history—history will later record that deliberations on the double celebration known in the novel under the heading of the "Parallelaktion" are already doomed to failure. No "real" importance accordingly attaches to the debates that occur throughout the long second part of book 1. But this is grist to the novelist's mill, as we have seen, since Musil directs attention not to action as such but to the kind of talk that is meant to anticipate it—Musil calls this part of the novel, not without

[7] I consider this important essay at greater length in chapter 4 and in the conclusion to this volume.

[8] A consideration of this chapter forms the centerpiece of the concluding chapter of this study.

irony, "Semblances Occur" (Seinesgleichen geschieht"). And here it again emerges that, as at the novel's beginning, life can be talked about, and experienced, from two markedly different vantage points: that of "precision," *Genauigkeit*, on the one hand, and that emanating in the feeling disposition of "soul" (*Seele*), on the other. At a key moment in the preparation of the "parallel action," the chief happening of the first book of the novel (in which, as just noted, nothing actually happens), the protagonist Ulrich, interrupting the chatter, steps forward to propose that "a general secretariat of precision and soul" be established.

The chatter must be imagined as proceeding at a swift pace. Nevertheless, Ulrich's reflective consciousness inserts itself into, and greatly enlarges, the interstices, as if normal time were suddenly suspended. We learn at one of these moments that Ulrich had lived his life in conscious knowledge of the existence of these two separate "trees of life": "His life grew from the separation of these two trees of his life" (*GW* 2:592).[9] We also learn that his decision to become a mathematician had arisen with some conscious intent that is linked to a strain of cruelty in his character (*GW* 2:591). A certain dissatisfaction with this strain of cruelty is implied, even as it is not wholly disapproved of. Dissatisfaction also attaches to the fact of separation itself—that the two trees of life cannot be fused through an act of willing consciousness, that the principle of "analogy" that animates the one cannot be made reliably to fall together with the other, the principle of "Eindeutigkeit" or Cartesian-style clarity, underpinning the progress of the mathematical and natural sciences of the other "half of life." And yet, as Ulrich's reflective consciousness also notes: "Without doubt, what one calls higher humanity is nothing other than an attempt to fuse together, once they have been carefully distinguished from each other, these two great halves of life, that of analogy and that of truth" (*GW* 2:593).

The union of opposites countenanced in this chapter and expressly suggested to the committee members of the "parallel action" by Ulrich himself is therefore far from a trivial matter. It might be imagined as an attempt to envisage a level of complementarity between elements that must otherwise be kept strictly apart—the kind of complementarity proposed by Niels Bohr in the 1920s in quantum theories that sought to generate a coherent scientific explanation for the wave-particle duality of atomic and subatomic matter. Ulrich's professional affiliation implies a sideways reference to this debate, which was occurring in scientific circles while Musil was writing the novel. The new demand Ulrich makes is that life should not only be lived but also be known, and perhaps also known *before* it is lived. It is the pure quality of this demand and the need

[9] The original German reads: "In diesen beiden Bäumen wuchs getrennt sein Leben."

for some compelling conceptual basis on which to advance it that brings Ulrich into conflict with a cultural environment that is too easily satisfied with unreflected living, or, as with the parallel action, too quick to duplicate the example of others. The encrusted cultural environment the novel examines is also too easily tolerant of received theories about how to carry out the business of living. Against this background, the reasons for Ulrich's "vacation from life" become fully evident. This vacation is designed to help him appraise what scientific knowledge makes available for the purposes of living, and, perhaps, to test whether this knowledge adequately serves the task of living.

In this important chapter in the novel, therefore, the two distinct languages of life experience introduced at the beginning of the novel run together. The protagonist is revealed over the course of the novel as borrowing heavily, but also equally, from both languages. He is also shown to be in crisis, precisely on account of the need to provide a convincing basis on which to allow both levels of life experience their due, without falling prey to the easy accommodations between these levels of experience that otherwise can be observed among those he mixes with in the Viennese intellectual and cultural elite of his time. These accommodations include ecstatic enthusiasm on the part of Clarisse, demagoguery on the part of the philosopher Meingast, slavish obedience to the passions for Ulrich's lover Bonadea, and, perhaps most insidiously, enshrinement of the principle of pure unreflected action in the case of Arnheim. Each of these accommodations comes up short, as the novel means to show. At any rate, Ulrich, the mouthpiece of the author's true feelings on the matter, is shown as opposing all of them on different grounds and finding no antidote to the crisis that has hit him in his thirty-third year of life. The crisis that afflicts him precipitates a crippling negativity toward those around him. At the same time, Ulrich's personal crisis suggests dimensions of a wider cultural crisis, precisely because these criticisms of others are so obviously prescient from the viewpoint of history. As Musil noted of the trajectory of the novel in his posthumous papers, "all lines [in the novel] lead to war" (alle Linien münden in den Krieg, *GW* 5:1851).

Ulrich's suggestion to found a "general secretariat for precision and soul" may be understood as a way of forestalling the entropic movement of affairs he alone appears to sense in Viennese society. His concern is to address the problem of how to link the two languages of being that have become vital for modern life. The novel may be read not only as highlighting the problem of these two languages but also as turning the need to find a basis for both to stand on into the actual source of metaphysical crisis in the novel. The novel's distinctiveness as a work of modernist fiction therefore comes in large measure from its attempt to consider whether scientific principles alone can be a legitimating basis, which is to say, whether the problem of the two cultures is answerable on the science side of the equation. If it were, the novel would effectively discharge the literary task foreshadowed in Hofmannsthal's *Chandos Letter* and, from a different viewpoint, the "antimetaphysical" trajectory outlined by the neo-Kantian philosopher and physicist Ernst Mach (on whom Musil wrote his doctoral dissertation). The novel thus envisages a vast thought experiment where the claim of science to legitimate the already beckoning technological modernity of current Viennese life is put to the test. Could there be, in other words, not just a "technical way of being" (*GW* 1:131), distilled from the essence of scientific procedure, but also a "legitimation of being on the basis of the technical," distilled from precisely this same essence? This question about whether science could offer an independent basis on which all disparate social commitments could be unified was to become one of the principal points of focus of the scientific group known as the "Vienna Circle." As the author's diaries record, Musil sat in on some of the meetings of this circle in the 1920s—the period in which the major novel took decisive shape.

## The Scientist and Scientific Ideas in *The Man Without Qualities*

A complex picture of Musil's use of science in the novel accordingly results. On the one hand, Musil's long-standing forward commitments are to the "analogical" and sometimes mystical realm of art, not the logical-analytic realm of science. On the other hand, his training as a scientist, which was deepened by his scientific studies in the first decade of the twentieth century, give him special insight into assessing the claim of science to constitute a basis on which to ground intersubjective commitments—whether these might be of an individual-ethical type, or a collective-societal type. Moreover, as Hofmannsthal's *Letter of Lord Chandos* makes clear, the new literature of Viennese modernism had already signaled an in-principle interest in these objective claims of science, as if art could somehow be moved forward to become a socially relevant and updatable, default science of progressive ethics.[10] Musil, too, clearly impressed by these claims, seems to accord some status to them in the novel. As mentioned, he even has the protagonist of his major novel float the idea of a "general secretariat of precision and soul," as if precision, from the realm of science, and soul, from art and religion, could somehow be reconciled. This idea initially appeals to Graf Leinsdorf, a leading presence at the gathering, but is rejected by Ulrich's adversary Arnheim as being naïve. No "general secretariat" is accordingly brought into being. Chapter 116 therefore marks the point where the ambition to forge a legitimate, which is to say,

[10] The ethical relation itself is touched on throughout this volume; it is prominently discussed in chapters 5–9 of part 2.

objective, foundation that unites the two languages of the world reaches a point of conclusion.

Before this point is reached, a fascinating dialog between art and science is unfolded in the novel. Over the course of nearly 700 pages, art and science, through a narrative technique that merges conventional narrative with its emphasis on character, plot, and dramatic action with an essayistic medium, with its emphasis on interpretation and analytical exegesis, execute a dance of proportions entirely original in world literature. Over much of this dance, neither the pretension of science to disclose the world objectively nor that of art to disclose it intuitively is ever quite allowed to carry the day. Ideas from both realms are brought into dialog with each other. The general tilt of events into war, for example, which, as mentioned, belongs to the narrative design, seems, at least notionally, paired with notions of entropy. As Musil's diaries indicate, he had good knowledge of the thermodynamic principles of physical matter developed in the 1920s by Erwin Schrödinger (see *TB* 1:524–25). From this angle, thinking of the novel from the viewpoint of science, its entropic design might well appear consistent with the second law of thermodynamics, a law which, applied to the open systems governing natural organisms (as opposed to the closed systems in which special conditions of local equilibrium obtain), suggests that entropy will always increase.[11] Even if this theory could be taken as a statement of a scientific principle that Musil is impressed by, is it consistent with what his science-facing art finally concludes about its own operations? Or is such a statement for Musil merely trivially true—part of the framing conditions within which all art that avails itself of the languages of life, including that of science, must operate?

At other times the properties of blending systems are introduced to suggest a similar conflation of science with art and nature with culture. Blending systems are systems in which, to quote Steven Pinker, "the properties of the combination lie *between* the properties of the elements, and the properties of the elements are lost in the average of the mixture" (Pinker 2015, 85). With blending systems, it is precisely the properties of the average of the mixture that are actively sought—in meteorology, for the purposes of assessing the weather, for example, or in paint mixing, in order to produce different shadings of color. The properties of the combination, however, obscure the properties of the individual elements. From this viewpoint, the discussion about the weather at the beginning of the novel is no mere conceit but anticipates a broader reflection in

---

[11] This appears to be Musil's understanding of Schrödinger's "laws of accidents" (Gesetze der Zufälle). As Musil observes in a diary entry: "That a continual progression of order to disorder results from this as a supreme law [*Obergesetz*], as the common basic character of events as a whole, seems completely intelligible" (*TB* 1:524).

the novel about the nature of systems whose properties are sought in the combination of different elements, not in the isolation of discrete elements. From this point of view, what constitutes "weather" is the combination of elements that produce the abstract measure of an average. We may arrive at something approximating this measure in terms of general sensation or intuitive feeling in ordinary life, or we may seek to identify it by following the procedural terms of scientific abstraction—say, the position of mercury on an abstract scale in response to thermal energy.

Musil clearly has something like a measure of scientific abstraction in mind when referring to "Kakania," his imaginary turn-of-the-century Viennese society. The social experiment Musil conducts in the first part of the major novel might be thought of, from this angle, as a combinatorial process whereby not individual being itself stages its appearance but only "Seinesgleichen," a "semblance" of "combinatorial" appearance of being that is not the same thing as true individual being itself. Following this insight, it might equally be possible to speak of the technical disappearance of the individual, who no longer appears under the guise of "individuation" (as in the traditional novel) but is henceforth transformed into a statistical measure. This measure would then appear as a calculation based on the sum of the properties making up all individuals. To this extent, then, the novel's interest in the blending systems of science in its portrait of "Kakania" leads to a provocation at the level of the individual. This is the assumption that humans can also be thought of in scientific terms—not as "undivided souls" with individuated being, whose occurrence is somehow unique in the world, but precisely as divided entities, as entities that are the result of combinatorial processes where being loses status as a singularity.

I want to finish these remarks by characterizing Musil's manner of using science in the novel, in narratological terms, as immensely productive. Yet it must also be pointed out that this use of science does not finally serve any truth discourse in the novel. The truth of narrative, which consists in sketching scenarios of subjective and intersubjective complexity, is ultimately at variance with scientific truth, whose interest is with analytical truth statements of a propositional nature. These latter propositions, indeed, are committed to eliminating the ground of narrative truth—the subject—from all consideration. Science is therefore formally committed to the cold impartiality that art, as a "hot" medium of partiality, emotion, and subjectivity, formally and by definition abjures. The suggestive blend of science with art in Musil's major novel—along with the crisis element that it also imparts, since the need for such a blend can be construed as a social demand in the period stretching from Ernst Mach and Hugo von Hofmannsthal to Husserl and the Vienna Circle—does not of itself involve a truth claim, for no "average" of the two in Pinker's sense obtains. Musil's protagonist breaks with the hope that a truth claim can emerge from the blend of science and art put forward in the first book of the novel. Instead, Musil strikes out in a new direction in the second book of the novel. The focus in this final part of the novel lies with a utopia—or, more accurately, a vast profusion of utopias—that Musil brought under the heading of the "Other Condition" (der andere Zustand). This is a condition where the descriptor "other" announces a search for a different kind of truth—a kind of truth that is "other" with respect to that which had characterized the imaginary society of "Kakania," where, as suggested, science could be offered as the technical ground on which to unify diverse social commitments and aspirations. The ground on which the utopias of this Other Condition are erected abandons the possibility of such unification, focusing instead on the ethical discussions that take place between two individuals, the protagonist and his sister.

This section of the novel, of course, was never finished. In its future focus, in its sense of the posterity it could only anticipate, perhaps we should say that it is unfinishable. What we have instead of a finished vision is a picture of the creative processes that animate art. The sometime Musil scholar and Nobel Prize-winning author J. M. Coetzee has this to say about these creative processes: "The feel of writing fiction is one of freedom, of irresponsibility, or better, of responsibility toward something that has not yet emerged, that lies somewhere at the end of the road" (Coetzee 1992, 246). This sense of responsibility toward what is yet to come, which Musil calls "Möglichkeitssinn" (sense of possibility) in the major novel, is the very antithesis of science, since the interest of science, as Wittgenstein put it, lies, in the final analysis, only with "what the case is," only what appears consistent with the factual order or the most scientifically legitimate assessment of it. While Wittgenstein enjoins us, from the viewpoint of science, to remain silent about what we cannot hope to know, Musil's unfinished final part of the novel encourages us in the feeling that we can never say enough about the state of our ethical selves and those ethical commitments we might hope to "finalize" in the future.

# 2: Musil, Ernst Mach, and the Problem of Causality

> *"There are objects," he thought, "below the horizon of consciousness. Objects that glide past, below the horizon of consciousness. Or actually just a curious, unfathomable, potential new horizon of consciousness, suddenly implicated, still devoid of objects."*
>
> —*Unions*

IN THIS CHAPTER I consider Musil's discussion of the philosophical problem of causality. This problem loomed large in Musil's doctoral dissertation on Ernst Mach, which he completed at the University of Berlin in 1908. Causality was a litmus test for the realism of the outer world. To question the validity, or at least the verifiability, of the causal connection was to weaken ontological claims about the status of the outside world. Hume—Mach's great forebear, in this respect—had famously posed seemingly unanswerable questions about the causal connection in his *Treatise on Human Nature* (1739), finding in that work that causal assignations were rarely, if ever, justified. In the absence of an unimpeachable standard for causality, Hume contended, one should more circumspectly talk of "constant conjunctions" between the elements of a causal relation. In matters of causality, then, Mach was self-evidently a Humean. But Mach also went beyond Hume, calling into question a great many things that appear to be indebted to this connection or make a clear investment in it: the notion of a world of substantial things, for one thing, but also the psychic agency, subjectivity, on the basis of which assumptions about that world are advanced. If causality could no longer be defended on strict scientific grounds, Mach suggested, then neither could the instance of subjectivity nor the subjects standing behind it. With conclusions such as these in view, Mach's antimetaphysical stance was well on its way to dissolving the subject-object continuum on which the substantiality of the world was considered to rest.

What follows in this chapter is close scrutiny of such ideas as we can fairly assume Musil to have received them. The weak realism of the Machian worldview, I argue in this chapter and throughout this study, was of great importance for Musil. It paralleled Musil's own sense of being sunk in an only notionally real world. At the same time, the Machian legacy clearly also worried Musil, for it undercut the realism to which science as a social enterprise was committed. For Musil, then, the Machian legacy cut two different ways. To be a scientific Machian was to be the sort of scientist Musil could never become—as Musil revealed through argument in his doctoral dissertation. Even if the urgings of Musil's supervisor Carl Stumpf, who seemed to cut Mach's philosophy little slack, can be discounted, Musil still felt himself compelled to find fault with Mach's scientific credentials. To be a literary Machian, by contrast, was to revel in a world of limitless possibilities, possibilities that were no longer held back by an ontological commitment to the outer world. To be a literary Machian, finally, was to ask whether the realism of science and the anti-realism of art and literature might somehow be resolved into some kind of third possibility. This chapter accordingly concludes with a reflection on what kind of third possibility this path between science and art might have involved for Musil.

In the next part of this chapter I look at Musil's responses to Mach in his doctoral dissertation. I begin this section of my analysis by considering some key features of Mach's thinking.

## Mach's Philosophy

The starting point in Mach's philosophy is the assumption that the physical and psychological realms are "essentially identical, intuitively known and factual, and only different on account of our perception of them" (Mach 1991, 13). Mach outlines a "psychology without soul" (12), which purports to describe the features of the physical world without recourse to objects that are deemed unobservable. His analysis dismisses these unobservable objects on the grounds that they generate only "pseudo-problems" (14)—problems that can only be approached speculatively, and therefore must be considered unfathomable and finally irrelevant for empirical analysis. From Mach's perspective, experimental methods must be applied if insight into the nature of the *Umgebungswelt*, the surrounding physical world, is to be obtained. But Mach's experimentalism is not satisfied with traditional empiricism. Mach considers the *Umgebungswelt* not to be composed of concrete objects at all, but rather *sentient values*—values he calls the "foundations of all psychological life" (24). Mach's empiricism restricts itself to describing the manifold "functional dependencies" that emerge in the *Umgebungswelt*, dependencies that are forever acted upon by new forces and which, in turn, constantly generate new formal appearances. While displacements and new associations thus arise, the totality of the world nevertheless remains roughly constant and stable throughout the process of change.

Mach's philosophy presupposes, accordingly, that natural phenomena are uniform and that a fundamental correspondence between "internal" and "external" processes is provided for in the notion of sentience.

Mach does not deny that the relational connections that arise from one moment to the next are perceived differently. He states only that the nature of a phenomenon depends on the prevailing sentient perspective. "Subject" and "object," "self" and the "world" are not verifiable concepts that have value for science, but are instead abstractions that fail to recognize that all experiences of reality are necessarily dependent—as we might say today—on how they are "mediated." On the other hand, such experiences can only be explored by means of a precise experimental methodology founded on the assumption that the spirit and the body, the subject and the self, represent different reference points for the same relation. Herrmann notes, in this context, that Mach's philosophy represents a comprehensive attempt "to discover in the physiological sphere an equivalent for every sensory-psychological item" (Herrmann 1982, 582). Conducting this kind of parallel undertaking, Herrmann believes, may be considered "more heuristically valuable" than an analysis of sensory data without reference to physiological circumstances (582).

Despite the methodological restriction to physiological circumstances, Musil finds in Mach's thinking in his dissertation a pervasive psychologizing of the physical realm. Musil says that this arises because Mach denies the concepts of subject and object their respective claims to substantiality. On one side, this abandonment of the parameters by which the nature of individual experience of reality is interpreted promotes an unrestricted alignment with totality; on the other side—and this is the psychological artifice of Mach's philosophy—this totality remains sensory-dependent. It is consequently not denoted by any operative regularity based on independent or absolute truth postulates, but rather on the unending variability of experiential structures, which are to be brought into harmony with the various respective manifestations of relational components. Since there is no substantive self and consequently no subjective perspective on conditions outside a subject, Mach's philosophy does away with the dualism of "internal" and "external" experience. Instead, Mach posits "a system of simultaneous dependent relations" (Musil 1980, 69) which can be subjected to precise empirical analysis and, in consequence, become observable for science. The crowning element of this relational analysis is its postulation of simultaneous correlations that need take account of neither temporal succession nor causal connections. The loss of succession and causality arises both because the relations do not occur in isolation, and also because (as we have seen) they must forfeit any prima facie claim to substantiality. As such, the immutability of space and time, which are the fundamental components of the causal relationship, and the claim to a priori knowledge, which presupposes that reality is knowable to the subjective viewpoint in principle on the basis of categories of understanding, are abolished. Instead of space and time, which are rejected as mere "constructs of thought," Mach identifies relations that accrue the status of a "practical convention conducive to their preservation" (Musil 1980, 34).[1] For this reason Mach replaces the causal relationship with the functional. Functional connections express themselves in mathematically measurable units (or, as Musil terms them, "determinate segments"[2]) and, when compared with the usually "interchangeable" concepts of "cause" and "effect" (Musil 1980, 69), stand outside time.

Two immediate advantages result from a functional understanding of the *Umgebungswelt.* First, the relations perceived by the senses occur in ever-changing forms, but are no less definitive for that. Rather, Mach says, when those relations are observed through a functional perspective they are in a position to attain completeness and precision. Secondly, the search for an ultimate cause as a matter of ontological concern for science can be abandoned, because there are no factors that will lead to success "at all times" (Musil 1980, 67). In other words, there is no physical necessity, and so no necessity inherent in objects themselves.[3] Instead of physical necessity, the door is thrown open in Mach's analysis to a new organizing principle, that of the progressive alignment of thoughts with facts (Mach 1991, 3). Anticipating the logical positivism of the Vienna Circle[4] of the 1920s, this economic principle displays a progressive encroachment of logical thought into the field of objectivity (2, 3). What refines itself and increasingly comes into view through analysis is thus not any object per se, but rather the operative economy of all thought itself. Mach's doctrines thus deny consciousness every "particular (psychic) quality or set of qualities that would differentiate these qualities from physical qualities" (43), since such qualities otherwise would require an elaborate definitional apparatus and, moreover, would need to proceed speculatively. Human volition is relativized with similar arguments: "There is no volition and no attentiveness reserved for specific psychological operations" (43). It is much more the case, according to Mach, that we forget too easily "how much [that occurs] runs mechanically in our own psychological

[1] The function of self-conservation is hereafter classified as the substantive project of science.

[2] In the original: "Bestimmungsstücke."

[3] See Nietzsche, whose critique of causality shares some similarities with that of Mach: "Necessity is not a matter of fact, but rather an interpretation": Nietzsche, *Der Wille zur Macht*, 1992, 391. Nietzsche calls the concept of causality a "deception" and disputes (as Mach does) the existence of the object per se: "An object is the sum of its effects, artificially bound together in a concept, a picture . . . In fact, science has emptied the notion of causality of all substance and kept the residue as an allegorical formula, in which it is effectively irrelevant, whether it is on the side of cause or effect" (389).

[4] It should be mentioned in passing that Musil did participate in the meetings of the Vienna Circle, but that his participation must have been "very modest" and remained "almost unnoticed": see Arvon 1970, 201.

life" (71). Similarly, he sees the difference between humans and animals not as qualitative but rather as "merely quantitative." The main difference between humans and animals that can be observed suggests that humans tend to formalize each advantageous process—they are, in Mach's eyes, "more economically inclined" (74, 76). On an operational level, economy gains expression in a principle of association, linked not to individual phenomena but rather to the great constancies of life: "In fact, acts of limited discretion by humans reveal themselves to be an equally definable regularity, like some kind of vegetative or even a mechanical process, in relation to which nobody ordinarily thinks about physical influences, or about the influence of volition" (28). Mach's doctrines, in the manner of positivism, thus restrict themselves to describing what is quantifiably measurable. Individual phenomena and "acts of limited discretion" can be removed from the analysis because they exhibit only variational tendencies with little or no bearing on the larger perspective *on their own*. Only by drawing larger correlations through empirical observation is the correct significance—characterized ultimately by compliance with rules—conferred on individual phenomena. Mach observes: "If, however, just one element without rules were a determining factor in relation to a mass phenomenon, no rule could emerge in the vast majority of cases" (28).

In the certainty that such an element is detectable, Mach's doctrines depart from acts that imply, or are openly premised upon, a substantial autonomy of the subject. Accordingly, there is no differentiation between qualities and quantities, no engagement with questions of being predicated on the concept of substance. Rather, Mach's philosophy sees its task as statistically determining the position of momentarily[5] arising relations of difference. Should any one determinative component prove to be unmeasurable, in accordance with the logic of the economic principle, it no longer belongs in the analysis.

## Musil's Reading of Mach

One of Musil's principal objections to Mach's thinking is that the question of causality receives no answer.[6] Moreover, the concept of function, which Mach defends in place of causality, imposes on phenomena a

[5] Mach applies a new principle, that of simultaneity, in order to make possible the calculation of determinative components (Mach 1991, 37).

[6] Bunge also finds Mach's rejection of the concept of causality problematic, because Mach's psycho-physiological parallelism, which is directed at the measurable units of the *Umgebungswelt*, achieves exactly the conditions for the reintroduction of the causal relationship: "despite Mach's abhorrence of causality, his account of mass is causal, because he assigns the very origin of that property to the environment—and, moreover, in a way reminiscent of Aristotelian dynamics, since the whole weight is put on the environment" (Bunge 1979, 175).

uniformity that cannot—or cannot always—be attested to. Musil proves this by reference to the production of heat through friction, a relation that, properly understood, maintains a clear distinction between cause and effect, before and after. The case of friction demonstrates, as Musil contends, that there are not always "simultaneous, invertible" correlations (Musil 1980, 87). In this case, a factual basis for concepts beyond their functional dependency is indicated, and, concomitantly, it must also be possible to make appeal to ontological considerations. From this standpoint, Mach's assertion that there is no physical necessity pertaining to objects can also be questioned—and for this reason: Mach's assertion would be invalid if it were possible to prove the existence of a set of facts on the basis of something other than the immediate effects of that set of facts on a person's sensate apparatus. In that case, we would be obliged to rehabilitate causality and see in the causal relation something more than Hume's constant conjunction (90).

Mach's universe comprises invariables, repetitions, and only minimal displacements. These displacements require no explanation, at most a more detailed description, or—as Musil says—a "disassembly into simplicity," "a restoration of familiarity" (93). The bottom line in Mach's thinking, according to Musil, is that there are no more surprises; nothing is capable of becoming a thing that does not already exist. In other words, Mach's model is a closed system.[7] The consistent correlation Mach observes does not imply reference to change or novelty in the natural world. Instead, it is only within the parameters of the preexisting system that change is possible. Moreover, such changes are experienced only as displacements within the totality, and those displacements are nevertheless subject to rules and do not impede lawfulness.[8] As discussed earlier, Mach rejects the position that in relation to mass-phenomena "even just one component without a rule would be determinative" (Mach 1991, 28). Thus it is clearly evident why Mach must oppose the concept of causality: causal relations rely on space and time, presuppose a factual (substantial) existence of individual components and direct attention to the problem of change in the natural world. "Internal" and "external" experiential values are observed together and given equal status. Causal relations retreat behind functional relations in this process. Henceforth, only those correlations are assumed that allow larger relationships of dependency to be perceived.

Musil correctly objects that Mach thereby engages in a form of psychology: ". . . everywhere the psychological viewpoint is encountered"

---

[7] Bunge considers that Mach's concept of functionalism does not allow for any mobility of the component parts; Mach's functionalism is more equivalent to the parts of an enormous network of metal scaffolding (91).

[8] For Nietzsche, the "predictability of an event" does not consist in the necessity of a process, nor in a law of causality, but rather in the "recurrence of identical cases": Nietzsche 1992, 390.

(Musil 1980, 102). In fact, a high degree of psychology underlies Mach's method of tracing experiential data back to an original economy of thought, which masquerades as the only inner necessity. The logical positivism of the 1920s subsequently expanded this kind of psychologism, culminating in Carnap's reductionist attempt to give all components of the experiential world the form of a logical premise.[9] The denial of physical necessity—one of the fundamental doctrines of Mach's philosophy—delivers objects over to a world dominated by modes of thought that can always be subjected to further, more stringent, analytical reductions. Musil observes that this process idealizes the elements it witnesses (132). Mach introduces conditions that ultimately approach an absolute idealism, though without thereby having to deal with questions of metaphysics. Musil asserts that Mach's doctrines are consequently founded on a "misunderstanding" (ibid.). Experience, Musil maintains, tells us that the regularity we recognize is found in facts, not in concepts. Yet it is Mach's guiding assumption that concepts always precede the facts (133).

We can now reach a summary assessment of Musil's responses to Mach.[10] The first criticism is that the analysis of experiential data of the *Umgebungswelt*, deduced through experimental methods, seeks to proceed empirically, but effectively places psychological categories at the center of analysis. The denial of the necessity of nature, which motivates such psychologism, is based on the assumption that facts follow concepts, and not the other way around. In Musil's view this assumption is misguided, because it is an error to assume that no substance inheres in facts, and that the regularity of facts (and therefore also their necessity), in consequence, is purely hypothetical. An explanation of the world of appearances is not exhausted by the description of dependent relationships that take their leave of facts,[11] nor is it contained in the discovery of logical principles inherent in perceptual values. On the contrary, the factual nature of the phenomena must be acceded to, and these lend themselves to causal analysis, as well as other types of analysis. If one reinstates causal relations, the objects of the *Umgebungswelt* become real again, and perceptions

[9] Scruton 1994, 201: "The hope was to define a relation of 'probabilification,' holding between propositions, and analogous to such logical relations as entailment and consistency, while admitting degrees."

[10] As Arvon argues, Musil effectively accuses Mach of "a lack of epistemological underpinnings," but acknowledges a method "thanks to which the real in its diverse and ever more profound layerings can be brought into view." Arvon is therefore of the opinion that Musil rejects only the "final metaphysical and epistemological conclusions (*Resultate*)" of Mach's doctrines (Musil 1980, 124), and is otherwise faithful to Mach's thinking (Arvon 1970, 205).

[11] Paul Feyerabend traces this loss of existence—which has been deliberately accepted by science—back to the inadequacy of abstract thought: Feyerabend 1995, 20.

are aligned with a really existing external world. The difference between "internal" and "external" experience must therefore also be reinstated. Mach's closed system, which does not permit any genuine novelty and is built on exclusively deductive principles, acknowledges in the manner of Berkeley only those internal conditions that can be perceived (which is to say, converted into an outward appearance). However, such a conversion is unidirectional, since the process of externalisation only reproduces what already exists on the inside. Mach expresses this in terms of a foundational economy of thought, the first principle of internality. Consequently, what at first glance seems empirical and objective has a decidedly solipsistic character on closer inspection, a conclusion that must be taken as a serious objection to the cogency of Mach's doctrines.

From this angle, then, Musil takes up a critical stance toward Mach in his dissertation. Musil questions Mach's thinking from the viewpoint of philosophical considerations. Musil finds that Mach is a scientist before he is a philosopher. Mach's science may well operate with degrees of success, but the philosophy he builds around it, at the very least, is faulty. As Musil's biographer Berghahn puts it: "Natural science and philosophy remain separate things—unfortunately" (Berghahn 1963, 56).

## The Influence of Mach in Musil's Early Works

Despite this philosophical rejection of Mach's empiricism in his dissertation, Mach commands a prominent his place in Musil's thinking from the very beginning. An early diary entry, which records Musil's first encounter with Mach's "popular scientific lectures," occurs during the drafting of Musil's first novel, *The Confusions of Young Törless.*[12] Thus even before the assessment of Mach was undertaken in his dissertation, Musil had identified in Mach "the presence of an intelligible set of teachings that were nevertheless of great importance" (*TB* 1, 20). Indeed, it would not be too far-fetched to look upon the protagonists of *Young Törless*, adolescent students at a military boarding school in Austria, as young Machians. These characters directly experience their *Umgebungswelt* in alignment with Mach's thinking and entirely without recourse to mature standards of judgment or prior sensitivity to moral precepts. It is precisely the absence of moral understanding that plunges young Törless into crisis when he is dragged into a campaign of brutality against a fellow student. Törless lacks arguments of his own to rein in or oppose the two ringleaders who instigate the campaign of violence. The "confusions" that beset him are only brought to an end when the violence, which in the meantime has spiralled out of control, comes to the attention of the school's authorities. Törless's discovery of the fact of the ethical relation, for this

[12] Arvon dates this journal entry as 26 February 1902 (203).

reason, is the real (though undeclared) conclusion of this story. Törless registers it with the finding that "things are things and will always remain so" (*GW* 6:138), a statement that must be taken as announcing opposition to Machian psychologism and support for a world in which there are two orders of being: the shifting order of the imagination, that revels in the constructs of the mind, on one side, and the more solid values of the received outer world, on which the standard of realism necessarily impinges and in whose soil the ethical relation is rooted. In *Young Törless*, then, the dualism of mind and substantial matter is reestablished.

In *Unions*, Musil's next work in terms of chronology, the pervasive dualism of the experiential world ratified at the end of *Young Törless* becomes the object of focus. Mach's influence in this work remains preponderant. If Musil had objected to the removal of causality as a structuring principle of understanding in his dissertation, *Unions* may be taken as evidence that he is far from wanting to reintroduce it. Instead, he appears to uphold a level of antipathy to direct causality and may even have contended that its application to the field of human emotions is entirely questionable, just as Mach had proposed. Musil's later observation in his diaries—"Refrain from doing anything causal, anything mechanical" (*TB* 2:700)—may be taken as further confirmation of Musil's abiding interest in psycho-physicalism, at least at the level of the emotions. If this insight is correct, Musil's relationship to Mach seems far less straightforward and far less uniformly condemnatory than the assessment of Mach's epistemology in Musil's dissertation might suggest.

It is precisely the preparedness to back-peddle in relation to question of causality that makes the project of *Unions* so beguiling. There appear to be two elements in play in this regard. On the one hand, Musil objects to attempts to subject sensate experience to the effect of causal laws. This clearly puts him at loggerheads with the project of logical positivism even before such a project was publicly advanced from the late 1910s and early 1920s. On the other hand, Musil also refuses to allow the spiritual dimension of humanity even the slightest regularity, and that seems bound up with his interest in defending the purview of the individual. The polemic in *Unions* against the type of rule-boundedness that the positivists will advocate is thus bought at the cost of a highly variable, if not openly idiosyncratic, subjectively premised picture of the world. The question of how the human soul and the world really influence each other, however, is left open in *Unions*. This leads to a strange set of results. On one side, the accent on the arbitrary quality of the spiritual dimension reinstates the terms of human inwardness. On the other, the abandonment of causal relations—relations on which ontological categories rely—has the effect of attenuating the substantiality of the subjective perspective and accentuating its unreality. A complicated trade-off is thus effected: subjectivity becomes both more autonomous, and yet, at the same time, more freewheeling, floating, and ephemeral. In intellectual terms, Musil's *Unions* project charts a movement in ideas from the scientific philosophizing of a Mach to the scientific polemic of a Nietzsche, yet without resolving the tension between the two standpoints (cf. Pieper).

In view of these circumstances, the following experimental intention can be said to underlie *Unions*: Musil follows Mach's polemic against causality, in order to achieve a rejection of the reductionist logic of science in regard to the emotions.[13] At the same time, Musil also appears to be in dispute with the functional method of observation of the "events" of the human soul. This is the case because no lawfulness in the emotions is ever identified in these stories. If functionalism has no currency for the emotions, the spiritual factor must arise by chance.[14] The measure of these spiritual factors can, under these circumstances, only be increasingly complex levels of sentiency developed out of the flux and flow of inwardness. Another way of interpreting the resulting crystallizations of energies on which human actions appear to be predicated in *Unions* is to view them as pre-linguistic imaginary pictures with "apperceptive" content. This would take away any sense of conformity to external laws, while also failing to subsume them under other connecting relationships for which a greater functional regularity might be claimed or required. These imaginary pictures would be compatible, at most, with positions argued by early Gestalt theory,[15] which attributes significance to individual components of a relation as well as to the fact of the relation itself.

In his diaries Musil uses the example of a melody to illustrate how Gestalt relations come about. As he observes, the melody is the "organizational moment"[16] affecting the component parts, which brings these

---

[13] Ulrich deplores the link in psychology between intelligibility and simplicity—the tendency to confer on a psychological thing, the simpler it is, more and more status as an explanation (*GW* 3:1146).

[14] Wolfgang Schraml has pointed out the significance of chance in Musil's work in his literary-anthropological study of Musil. Of primary importance in this context is his reference to the question of causality, which has been extensively misunderstood in research on Musil, but which he admittedly does not take any further (1994, 193–94).

[15] Musil probably became familiar with the origins of Gestalt theory during his studies in Berlin with Carl Stumpf, his supervisor: see Arvon 1970, 202.

[16] Roger Scruton states that the aspect of organization is present in all perceptive activity (Scruton 1994, 342). The absence of this cognitive organizational ability, called "aphasia" in clinical psychiatry, is described for example in Oliver Sacks's case study in *The Man Who Mistook His Wife for a Hat*. In *The Man Without Qualities*, Musil's character study of a man of unsound mind (Moosbrugger) who rapes and murders a prostitute similarly results from his interest in holistic theories and further demonstrates that Musil was ahead of his time in his description of aphasia. NB: An early form of Gestalt theory can be found in the work of Leibniz, particularly in the context of his monistic conception of the soul (of some significance for Musil): cf. Specht 1990, 250–51.

parts under a totality.[17] In regard to the problem of how these relations are to be conceptualized, Musil then states: "The grounded content along with its attendant individual perceptions is the complexity we then become aware of" (*TB* 1:131). It is precisely the process by which awareness of such complexity is attained that Musil wants to make evident in *Unions*. We may therefore look upon Musil's concept of Gestalt as an attempt to maneuvre in the space between Mach and Nietzsche. Gestalt formations, from this angle, represent instantiations of the component parts in a schema of rising complexity. Musil will refer to them in the later major novel as "a special form of existence" (*GW* 3:1164). These instantiations both reintroduce the category of substance and, against the presentiment of conventional philosophy, repudiate the causal relation.[18] Musil thus deliberately goes beyond a purely psychological application of the concept[19] as it was advanced by the early Gestalt theorists Ehrenfels and Meinong as well as the later so-called Berlin School.

The two principles on which *Unions* is constructed—the polemic against causal relations and the primacy of the spiritual dimension—can be understood as vital impulses running throughout Musil's work. They remained with Musil long after critics had declared *Unions* to be a literary failure. That *Unions* proved incomprehensible for Musil's audience is partly attributable to the difficulty of the intellectual project he was aiming at. He was clearly also ahead of his time in thinking about how the methodological commitment of science to precision might be paired with art and literature, which sought to reserve for human understanding greater freedom in view of a new awareness of the problem of human complexity—the quality Musil appears most committed to in *Unions*.[20]

We see both these aspects at work in Musil's later fiction. In the *Three Women* project, for example, none of the focalizing figures of each of

---

[17] Cf. *The Man Without Qualities*, in which the thesis is advanced that a melody does not develop next to the notes, tonal intervals and timings, but rather *with* them: "The melody is not added as an accompanying element (*Beigabe*), but as a second way of appearing, a particular form of existence under the terms of which the individual form of existence can only be removed in a limited sense (*unter der sich die Form der Einzelexistenz gerade noch ausnehmen läßt*)" (*MoE*, 1164). In the context of Ulrich's thoughts about the form of melody, Honold has drawn attention to the "paradoxical determination of time," which results from the dual modality of separation and connection (Honold 1993, 314).

[18] Luserke similarly finds in Musil's concept of Gestalt a "reference to being," which can then be placed in a further relationship to the concept of possibility (Luserke 1987, 267).

[19] This fact is emphasised in one of Musil's journal entries: "I sought [in *Unions*] the true (ethical not merely psychological) determinants of action" (*TB* 1:232).

[20] See the concluding chapter of this volume for discussion of Musil's mature statement of this problem in *The Man Without Qualities*.

these three stories is served well by adherence to science or to the type of rationalist principles on which science is predicated. In the third story of the volume, "Tonka," not only are doubts raised against reason as a general standard by which to orient behavior, but the principle of causality, the very touchstone of such rationality, is openly questioned.[21] The arc of logic countenanced in this story, from this angle, does not lead in the direction of scientific reason but, on the contrary, toward ratification of the need to exercise a form of trust toward another human being in conscious opposition to the scientific viewpoint (note that the central character in the story is a scientist).[22] The second story in this collection of stories, "The Portuguese Woman," brings this insight into even sharper relief. The critical moment in the protagonist's life takes place when the warrior's lust for battle suddenly weakens after a period of hostilities in which he has again been victorious. The warrior's logic, which rests on a merciless, calculating, mechanical causality ("One action takes place because another has occurred"[23] [*GW* 6:256]), is put to question shortly before he returns home. He is laid low after suffering a seemingly harmless insect bite. The resulting infection quickly engulfs his body.[24] Yet acquaintance with the scientific materialism of the clinic does not rehabilitate any order of rationality in the story. Moreover, the illness afflicting the protagonist seems to imply a disease of inner states.[25] Only a final act of extreme recklessness launched from within, precisely addressing the misalignment of inner states with the physical body, provides deliverance for the main character. When Lord von Ketten climbs the outer fortress walls (where to miss a finger- or foothold would be to plunge to his death), improbably survives the ordeal, and returns to his former life and to his lover, the beautiful Portuguese woman of the story's title, he not only shakes off the bonds of his lineage and its rigid adherence to a primal state of warfare but also cuts through the simple order of the spirit

[21] Schraml saw in the stories "Tonka," "The Portuguese Woman," and *The Blackbird* the "literary impact of Musil's preoccupation with mathematics and statistics" in regard to which the capacity of chance to "[reduce] the validity of causal statements and [cancel] out all deterministic assignations" becomes central (Schraml 1994, 195).

[22] See Diary Notebook 8 from 1920: "The legitimate measures of cognitive reason (that which can be objectively measured and legitimized) are applied illegitimately to the entire inner life (*seelische Leben*) [of the person]" (*TB* 1:362).

[23] "Tat geschieht, weil andre Tat geschehen ist."

[24] According to classical (causally understood) processes of illness, any active internal involvement of a sick person in their own sickness is excluded. As such, the sick person is understood to be a "patient" who "awaits" their recovery in accordance with the meaning alluded to by this term (Bunge 24, 179).

[25] As Bunge puts it in a philosophical context, "much of what is true of infections is also true of affections" (Bunge 179–80).

by which he has lived till now. Von Ketten literally "breaks the chains" of dependence on the order of mechanical causality.[26]

Musil's award-winning drama of 1921, *The Enthusiasts*, grapples again with the Machian problem of functionalism, ten years after its first extended reflection in the *Unions.* Stader, a private detective engaged to shed light on a case of adultery in the play, seems to preach directly from the Machian songbook, parroting a passage from Mach's *Erkenntnis und Irrtum* when he observes: "In summer there are more pregnancies, in winter more suicides" (*GW* 6:394).[27] A later remark echoes Mach's rejection of dualism: "There are no facts—to be sure! There are only—scientific relations" (*GW* 6:339). Mach's theories are brought into literature in this way, but not with the intention of fictionalizing them or rendering them harmless and effete. Rather, they serve Musil's intention of asking basic questions about the meaning of human existence at a time of deep crisis about the value of that existence. These questions come into view in the play when the scientist Thomas asks: "Where is soul, order, the law of the spirit? Belonging together, being comprehended, grasping something? Truth, real feeling?"[28] (*GW* 6:367) From Thomas's viewpoint, these are open and difficult questions. The Machian standpoint, however, would suggest that they do not require an answer. Anselm, a second mouthpiece for Mach in the play, registers this viewpoint in a way that recalls Mach's basic functionalism: "Even something out of the ordinary is nothing more than a reversal of the ordinary"[29] (*GW* 6:345). Where Thomas directs feelings through conscious calculation in order to find his identity, Anselm trusts that the powerful pull of raw emotions on their own will be sufficient to give him direction. Where Thomas locates in individuality the seat of all perception, Anselm thinks that a larger view of the logic of the emotions will allow the individual to find his true place in relation to others.

The basic conflict explored in *The Enthusiasts* thus looks like this: if humans turn off their intellect in order to access their true inner being, they put themselves onto a precarious path. What starts out celebrating

---

[26] As Musil states in his journal: "The causal chains of human development and those of the life form (*Lebensform*) are different" (*TB* 1:541).

[27] See Mach 1991, 28: "The number of annual marriages and suicides in a particular country fluctuate just as little, or even less, than the number of births and deaths by natural causes, even though, in the first case, the will is in question, whereas in the latter case it is not." See also Musil's excerpts in his journal, diary volume 19 (1919–21) from the statistical yearbook for the German Empire (34th year, 1913), in which inter alia suicides are specified for the year 1911 (*TB* 1:535).

[28] "Wo ist Seele, Ordnung, geistiges Gesetz? Zusammengehören, Begriffenwerden, Ergreifen? Wahrheit, wirkliches Gefühl?"

[29] "Auch das ungewöhnliche Erlebnis ist nichts als eine umgestülpte Gewöhnlichkeit."

the power of the emotions ends up having very little to do with the individuals themselves. In that case, as Thomas's sister Regine, formerly Anselm's lover, points out: "When one no longer has the strength to be different from the actions one carries out then one is no longer a human being!"[30] (*GW* 6:348) The conflict of the emotions treated in the play, to this extent, emerges from the application of Mach's doctrines. When Regine traces the flourishes and circles of a carpet pattern with her feet, she is led to an important insight: "I can move everywhere and yet not move anywhere"[31] (*GW* 6:402). The insight seems directed against Mach's determinism, whose commitment to regularity provides security but imperils the freedom of the individual. The path to the truth of inner feeling, for this reason, must provide for the fundamental dualism of the world. Accordingly, Musil appears to settle on a position that rejects Mach's parallelism, under which physiological and psychological realms are merely different appearances of the same underlying relations. Whether, beyond the dualism of the world established in *Unions*, a third way might nevertheless be discovered—a two-sided aspect in understanding, under which the physical and spiritual could be understood as two sides of an "unknown third side" (Hermann 582)—is a consideration whose treatment is postponed in Musil's work and thought to the major novel, in particular to book 2 of the novel.

## Mach and *The Man Without Qualities*

Musil prepares for this third possibility by once more taking up and radicalizing key premises arising from Mach's doctrines. The descriptor "the man without qualities"—seen from this perspective—can be read as an intensification of one of Mach's key claims. Musil's conceit is consistent with Mach's general position that subjects and selves are not interesting on their own but only gain significance, albeit with different effects, when viewed from the perspective of the basic conditions of life at a macro level of analysis under the rule of the principle of economy. The individual, according to this view, is not a discrete and autonomous part of the whole, aggregated with other parts to constitute a whole. Instead, the individual, to the extent that individuality can be spoken of, is constituted as a part-aspect of the whole. From this angle, the whole supervenes on the part and furnishes the part with an identity conferred by its articulation with the whole; the part has no identity separate from the whole, and thus no meaningful autonomy. As with the three-dimensional

---

30 "Wenn man nicht mehr die Kraft hat, etwas anderes zu sein, als man tut, ist man kein Mensch mehr!"

31 Ich kann mich überall hin bewegen und kann mich doch nicht überall hin bewegen."

photocopier from our own day, which works only from a conception of the whole and no longer that of the part, the part is never constituted as a part but only "appears" when the whole is fully constituted. Musil expresses this thought in his diaries—now in reference to human individuality—in the following way:

> Human beings have as it were accidental attributes [*Eigenschaften*], side reactions that influence their fate but do not have anything to do with the human beings themselves.—These attributes are attached to human beings only loosely, they never reflect on why or for what purpose they have them, they are in fact quite foreign and indifferent to the human being, even though they oftentimes determine that human being's outward fate and somehow even causally participate in the construction of their being.[32]

Causal relationships—under the assumption of a now more inert causality—no longer describe an autonomous being that can be separated from the whole.[33] The component part, the human individual, is instead "washed to the edges" (*GW* 3:907). Under these circumstances, the individual is no longer indivisible; indeed, one can no longer properly speak of an individual in the old way.

All significant conceptual aspects of the literary program of Musil's major novel can be traced to the thought experiment of *Eigenschaftslosigkeit*, the condition of being without attributes or qualities. To take one example: the notion of the "absence of qualities" implies a distinction between a "person" and a "citizen," because a citizen comprises the aggregate of attributes that participation in social life presupposes, whereas in the case of a person attributes do not come from the prior fact of society but are inherent aspects of what makes up a person.[34] Accord-

[32] "Menschen haben viele sozusagen zufällige Eigenschaften, Nebenreaktionen, die ihr Schicksal beeinflussen können, mit ihnen selbst aber nicht zu tun haben. - - Diese Eigenschaften hängen viel lockerer an ihm, er hat nie darüber nachgedacht, warum und wozu er sie hat, sie sind ihm eigentlich fremd u. gleichgültig, obwohl sie oft sein äußeres Schicksal bestimmen u. irgendwie sogar causal an dem Aufbau seines Wesentlichen beteiligt sind" (*TB* 2:1103: Anhang: Heft 8 [1920]).

[33] See *GW* 3:907: "'But the older one gets, the clearer one sees that this apparent superfluity, this autonomy and mobility in everything, this independence of movement of the parts (*Souveränität der treibenden Teile*) and the movements of the parts (*Teilantriebe*) is nothing other than a weakness of the whole in relation to its parts. . .'"

[34] Jochen Schmidt's indication of the significance of Meister Eckhart's philosophy in understanding the concept of "the absence of qualities" equally targets a differentiation between socially mediated identity, which is reflected in the worldly qualities of a person, and the human essence that precedes this identity. However, his emphasis on the mystic alignment with God appears to be one-sided: (Schmidt

ingly, in Musil's understanding, a person remains always other than her attributes, since attributes are derived from the participation of society in the "construction of [her] being."[35] Nevertheless, it is none other than Mach's functionalism that provides the best means to describe the new kind of society Musil examines, a society that the moniker "Kakania" suggests is a hollowed-out society, and one soon to fail. Musil invents the phrase "semblances occur" ("seinesgleichen geschieht") in order to characterize the frenzy of repetitions and imitations that characterize this society—a society whose chief concern is expressed in a "parallel action" by which the impending jubilee celebration of the Austrian monarch and the German emperor's jubilee due in the same year[36] are to be linked. The human subject, from this angle, is brought into conflict with the logic of emerging mass society. Whether that subject can be isolated from this logic and recovered from its encroachments is one of the chief questions the novel poses.

When we examine the contribution the philosophy of Mach makes to Musil's portrait of turn-of-the-century Austrian society in *The Man Without Qualities*, therefore, we must conclude it to be immense. The conceit of *Eigenschaftslosigkeit*, the idea of the "parallel action" by which two state celebrations are to be conjoined, the notion that semblances might be seen to "occur" in blatant disregard for the substances lying somewhere behind them—these are propositions or conceits Musil derives from Mach and turns to literary effect. To the extent that Mach's functionalism, to which these propositions ultimately owe allegiance, is—or means to be taken as—a way of doing science, Musil's project in the *The Man Without Qualities* involves him in a critique of science. The radical nature of the major novel lies with the proposition that it is possible to bring the abstractions and technical innovations of the scientific enterprise into the realm of literature, where they can be evaluated as social ideas. Musil's novel thus requires of us a view about the plausibility of this use of science. What is the relation, then, of the abstractions of science to the practice of daily life? How are we to avail ourselves of these abstractions, and where should the line be drawn between the conduct of science

1975, 46–53). As Schraml observes (1994, 79–80), Manfred Frank takes this fundamental idea further and establishes a connection to Mach's monism, albeit without acknowledging the consequently problematic question of the substancelessness of humanity, which represents one of the significant aspects of tension in Musil's thinking (Frank 1988, 315–32).

[35] In Musil's journal the same argument appears: "The meaning of the human individual in social life would have to reside in a type of being of his soul for which [individual] actions (*Leistung*) remain unimportant [*eine Äußerlichkeit*]" (*TB* 1:363).

[36] 1918—five years hence in terms of the novel's temporal setting.

and the needs of practical life? How—in view of the edifice of science that rises ever more steeply before us—are we to live?

While no clear answer seems to be given in the novel to these questions, it is evident that Musil considered it possible, at least in theory, to provide a plausible response to them. The response to Mach's thinking involves neither ratification of the ends of science nor condemnation of them. Rather, it points to the possibility of a third way,[37] the possibility of allowing science a role in the attempt to define, with ever greater clarity, the terms within which life must be lived, while simultaneously reserving for the non-scientific attitude—the attachment to the terms of life that Mach's thinking expressly excludes, terms such as subjectivity, soul, dreams, the imaginative constructions of being—an equally significant role. This third possibility emerges on account of the impoverishment that inevitably occurs when either one of what Musil calls the "two trees" of modern life is downgraded or eliminated: "According to the manner of bacterial strains which divide the organic into two parts, the human strain breaks the original state of the life of images into the firm matter of reality and truth on one side and the glass-like atmosphere of intuition, belief and artifice on the other. There seems to be no third possibility between the two. . ." (*GW* 2:582).[38]

So where might this third possibility be found? Musil's philosophy of feeling, which is developed in the second book of the major novel but already anticipated in the closing chapters of the first book, provides the beginning of an answer. For one thing, feelings oppose the sequentialism of the causal relationship, a determinant of purely physical life. But this sequentialism—the "immediately next thing" ("das Nächstliegende"; *GW* 2:581–82), in Musil's parlance—has no bearing on moral life, and it is this aspect, Musil thinks, in regard to which we should remain vigilant. This was already the lesson arising from *The Confusions of Young Törless*—purely physical life is devoid of moral bearings and cannot on its own develop them. The rejection of causality, then, marks a separation between physical life and moral life and is equally a reason for the focus in later works on the inner life of the emotions.[39]

---

37 The first mention of the possibility of an as yet undiscovered "third way" is on page 24 of Musil's diaries in 1904–5: "The question that must be put to start with is: is logic to be approached psychologically? Or by way of metaphysics (*metaphysisch*)? Or is there a third possibility?" (*TB* 1:127).

38 "Nach Art jener Bakterienstämme, die etwas Organisches in zwei Teile spalten, zerlebt der Menschenstamm den ursprünglichen Lebenszustand des Gleichnisses in die feste Materie der Wirklichkeit und Wahrheit und in die glasige Atmosphäre von Ahnung, Glaube und Künstlichkeit. Es scheint, daß es dazwischen keine dritte Möglichkeit gibt . . ." (*GW* 2:582).

39 It is this knowledge, which registers the impending flattening of life, against which Musil's mistrust of "the doings of the body (*das körperliche Treiben*)" (*GW*

In the final part of the major novel Musil aims to construct a utopia of emotions—or many such utopias. These are built from the conception of a love that "has no causality and knows no lawfulness" (*GW* 3:1127). The emotions Musil describes are conceived of as events that bring the entire life of the person into their orbit (*GW* 3:1158). When these emotional events bring about change, it is not with reference to any sequentialism or lawfulness, but rather involves a process of "reciprocal dependence and requirement" (*GW* 3:1159) between emotional factors. But this should not be taken as suggesting that emotional complexes stand outside time (as Machian functionalism does). On the contrary, emotions, to Musil's way of thinking, occasion palpable effects in real time—effects that go beyond the person who experiences them. Ulrich, speaking for the author Musil, expresses it in this manner: "A feeling I call my own forms in me and outside of me; it changes me from the inside and the outside; it changes the world directly from the inside and also does so indirectly from the outside, through my behavior" (*GW* 3:1161).[40]

Yet if we were to seek to understand the respective proportionality of inside to outside in a given feeling, we would find no answer. This is not least because the attention required to be directed at the emotions in order to provide such an answer would already bring about a change in the state of the emotion we seek to understand and analyze (cf. *GW* 3:1172). We are obliged to accept, then, that emotions have substance, but this substance eludes direct physical observation. Emotions exist factually; they arise in the corporal-spiritual duality of human experience, and thus have a physical relation to the world, which also means that they are more than mere psychological aspects of a vast cosmic logic. Emotions thus depend on the person (where they are experienced holistically and internally), and on the other hand are rooted "as a property in things and events" (*GW* 3:1162) in the external world.

The classifications that had been rejected in Mach's doctrines, viz. the indubitable reality of experience as well as the paths of freedom by which emotions and the world intermingle, are thus rehabilitated. To this extent, Musil does not finally imagine humanity existing as a "thread in the vast weave of a world-material," but rather in the manner of a person "who stands on this carpet of threads" (*GW* 3:1144). Musil's engagement with Mach's doctrines, to which this image from one of the last chapters

1:30; 183) of his time can be assessed. The achievements of athletes can be measured perfectly (*GW* 1:45), Musil regrets, whereas ideas, which develop within people, first take hold on the surfaces of the body. An "age of body culture" seems to thereby be brought about (*GW* 2:380).

[40] "Mein Gefühl bildet sich in mir und außer mir; es verändert mich von innen und von außen; es verändert die Welt unmittelbar von innen und tut es mittelbar, das heißt durch mein Verhalten, von außen [. . .]"

of the novel recurs,[41] thus reaches a point of conclusion at a vast remove from its point of departure. The type of life to which the *The Man Without Qualities* ultimately directs attention is rich and complex; the type of subjectivity bound to that life engages with it "directly and undividedly" (*GW* 3:1162), forever anticipating an idea of complete experience that always eludes it.

[41] The so-called type-set chapters, which Musil eventually withdrew before publication.

# 3: Musil's Theory of Vision

In an essay published in 2007 in the journal *Seminar*, Rolf Renner outlined a theory of "intermediality" in which similarities discernible in forms of art outweigh any intrinsic differences between them (Renner 2007). In its philosophical assumptions, this discussion runs counter to the direction of Lessing's project in his renowned essay *Laocoön* (1766), insofar as Lessing considers ostensible discrepancies between sculpture and poetry in the rendering of the time-space continuum to be of sufficient merit to warrant a basic distinction in terms of method and outlook between the two art forms. Lessing in this essay formulates the position that temporal references are not only treated differently in poetry, but in fact have greater significance and value in this medium than they do for the visual arts, whose greater interest is in the spatial dimension and the organization of space. For Lessing, responding to Winckelmann's reflections on the art of antiquity, "intermedial" difference is of high importance and affects the kind of messages we derive from the various art forms. To the classical Laocoön sculpture, for example, Winckelmann had attached a principle of restraint—what he called "noble simplicity and quiet grandeur."[1] In Lessing's view, it was the nature of the visual medium and the constraints imposed by it that led to such a view, since it would have been unseemly for Laocoön or his sons to express the pain inflicted on them by the serpent directly. By contrast, for the "temporal art" of poetry, the Roman poet Vergil in the *Aeneid* throws caution to the wind in conveying the suffering of Laocoön. Vergil's Laocoön, in the medium of poetry, gives the emotions free range.

Critical responses to the Laocoön statue group remind us of the importance it possesses for humanist discourses in the early modern period. Whether Laocoön yields to the pain he suffers or whether he rises above it is of great interest, but of even greater significance is how the various art forms treat the subject of pain. On the eve of the emergence of the modern in the era of the late Enlightenment, pain is a "neuralgic" point requiring its depictions to be differentially allocated. That the modern will depart from the logic of this allocation is clear from Renner's assumptions. Renner says intermedial differences have been exaggerated. For Renner, focusing on the issue of the depiction of space,

[1] See J. J. Winckelmann, *Gedanken über die Nachahmung der griechischen Werke in der Malerei und Bildhauerkunst.*

all media, whether electronic or paper-based, visual or lettered, conventional or experimental, draw attention to space in roughly similar ways and indeed are confronted with the same question of spatial simulation despite their diverse manipulations of themes and motives and their particular materiality. Extending from this argument about space we can say, following Renner, that all media strive to create and cultivate similar basic effects. To add weight to his point, Renner cites illusionary effects, which in regard to the graphic arts are defined as the ability of the artist to overcome the technical and material constraints of two-dimensionality through the simulation of an idea of three-dimensionality (Renner 386). In accordance with this proposition, every artwork is obliged to apply illusory techniques in order to allow the imaginative presentation of space to develop in effective ways for its audience. By discussing the importance of the relationship between the individual media, Renner avoids the sort of inherited demarcations within art and between the various art forms derived from Lessing. Thus emerges a complex, cross-disciplinary dimension of spatial simulation, which may be seen as a technical task of fundamental concern to all forms of art.

Renner's observations in relation to the aesthetic construction of space in art offer a productive starting point for such questions in relation to Musil. In the short text "The Flypaper," Musil breaks with the foundational humanistic assumptions of a Lessing or Winckelmann, finding the physical exertions of the fly more true-to-life in its "sporting expression" than the struggle of Laocoön and his sons. If more truth attaches to what we witness of a fly trapped on flypaper than to the "silent grandeur" of the Laocoön group, then, at the very least, our humanism is less vaunted, and perhaps also less nobly composed, than we think. Conclusions about a "posthumanist" world based on a differently constructed hierarchy of life-forms than those seen from the humanistic standpoint would appear justified. If Musil's discussion in "The Flypaper" permits these types of conclusions and allows us to see at least a prefiguring of posthumanist thought, does Musil's understanding of the relations between the various art forms conform to Renner's views on the matter, namely that time and space are shared attributes of an intermedially operative art, not allocated to the various art forms according to the requirements of these forms? To advance discussion of this question, I now turn to Musil's short prose work "Trieder," one of the texts included in Musil's last published work, *Posthumous Papers of a Living Author* (1936).

## "Triëder" and "Technical Being"

As Birgit Nübel argues, Musil's striking interest in literary experimentation appears to be indebted to certain techniques in film, the newly emerging art form of Musil's day. In this context, Nübel understands

Musil's "Triëder" to be a text that seeks to imitate a filmic procedure of "decontextualization" and "depragmatization" in relation to the standpoint of the observer (Nübel 2006, 481). These decontextualizing and depragmatizing effects result from Musil's attention in "Triëder" to the mechanism of isolation. The program Musil outlines in this short piece is premised on the argument that every object can be seen, at least initially, as an object in its own right, that is, as an object that stands out from other objects it appears with in the wider situative contexts in which it is found. Such programmatic depragmatization of vision also occurs in film, most obviously with film techniques such as the "close-up," which, as Béla Balázs discusses in *Der sichtbare Mensch* (1924), indeed appear to reveal a common intention to make the human face of the modern age properly visible in its true form for the first time. For Balázs, admittedly, this depragmatizing technique anticipates a "repragmatization" focused on a return to the old humanistic educational ideals of the pre-industrial era. Musil knew Balázs's work. He was also involved in the reviewing of films during his time as newspaper critic and clearly spent some time thinking about film as a medium.

While Musil was clearly interested in the general reappraisal of the terms of vision in common with the new medium of film, this reappraisal—as I explain in this chapter—was not ultimately derived from filmic sources. What Musil values about film is the technology that underlies it. This understanding of the importance of technical devices and contrivances, for example, is missing from Nübel's assessment, which focuses on the experimental effects Musil was interested in, and which are also encountered in filmic technique. When we shine a light on the question of technology and the technical in Musil, however, we get a different "take" on the question of these effects and what Musil understood about their importance—one that leads us again back to questions of science in Musil's outlook. What weighs more in Musil's thinking are insights he draws from scientific sources and his developing view about problems and opportunities that lie in the space between the two cultures. Musil's reflections on these problems and opportunities, as has already been foreshadowed, are a major concern in *The Man Without Qualities*, where terms such as "technical being" (*Technik des Seins*) are glosses for this interest and the problems associated with it. The short work "Triëder" is valuable, I believe, because it concentrates interest on the question of what Musil might have intended with terms such as this. Did Musil believe that science and technology could lead to new and better ways of living, or was he simply aware of intellectual currents circulating in his era, which, in view of their capacity to use the technical to go beyond the human, would ultimately derail older projects such as the humanist? Is Musil, in short, ultimately to be considered a humanist who was worried about the encroachments of what we now call the "posthuman," or was he an early "posthumanist" who saw promise in the capacity of a technology and "technical being" to solve the social and psychological problems that humanism was unable to solve?

An early expression of these questions can be found in Musil's first novel, *The Confusions of Young Törless*, which Musil began in 1903–4 and published in 1906. The young boarding-school student Törless, who is exercised by a forensic interest in the violent acts of two classmates, without fully participating in these acts himself, is not able to come to a clear moral position in relation to the brutality of the students. In light of the "confusions" to which he is led in the course of the novel, and which he cannot dispel despite repeated efforts, he comes to this conclusion: "I know: things are things and will always remain so, and I will view them this way and that. Sometimes with the eye of rational understanding, sometimes with another eye . . ." (*GW* 6:138).[2] These words uttered toward the end of the novel indicate that the narrator and the author Musil know that there are two kinds of human understanding located in different experiential complexes available to humans—technical, abstract and "neutral" understanding, associated with science, and emotional evaluative moral-critical understanding, associated with art—that ultimately cannot be brought into harmony with each other. As I discuss in this chapter, Musil appears to arrive at a more tolerant position in later works such as "Triëder." Nevertheless, we already see that *The Confusions of Young Törless* outlines the nature of a problem that all Musil's later writing draws from and ruminates upon: what do science and scientific ways of seeing the world contribute to the general cause of understanding? How should ordinary understanding take account of the insights of science and the progress it achieves? As Hoheisel notes of Musil's work in general: "Musil's literary work is scarcely comprehensible without at least a basic understanding of the natural sciences, mathematics, and statistics. The reason for this lies with the direct way in which these special fields are operationally tied to his work. [. . .] The connection between the various areas of knowledge (the scientific, the mathematical) is never achieved merely as a matter of metaphor or the construction of analogies, but requires that the scientific discourses [and art] be viewed together" (Hoheisel, 462).[3]

---

[2] "Ich weiß: die Dinge sind die Dinge und werden es wohl immer bleiben; und ich werde sie wohl immer bald so, bald so ansehen. Bald mit den Augen des Verstandes, bald mit den anderen [. . .]" (*GW* 6:138).

[3] "Musils literarisches Werk ist ohne elementares Lehrbuchwissen der naturwissenschaftlichen und mathematisch-statistischen Spezialdiskurse nicht adäquat zu lesen. Der Grund liegt in der direkten, operationalen Ankopplung der verarbeiteten Fachdiskurse. [. . .] Die Verbindungen zwischen den verschiedenen (naturwissenschaftlichen, mathematischen) Wissensbereichen werden bei Musil niemals

## Education

It is noteworthy in Musil's literary development that from the very beginning he had a strong, and indeed intimate, relationship with technical science. He grew up in a technically inclined family. His father Alfred, a qualified engineer, became a professor of mechanical engineering at the Technical University in Brünn (now the Brno University of Technology). After a short stint at a military academy in Vienna in 1897, Musil junior took on three years of study in his father's department in Brünn, and then completed a one-year elective under Professor Julius Carl von Bach at the Technical University of Stuttgart. From this engagement with technical science we can assume that Musil's professional path was expected to lie more in the direction of technology and technical science than literature. Musil's earliest literary activities, indeed, were conducted—in line with this assumption—in his after-hours time, after completing work in the laboratory during his elective year (for a more detailed account of this period, see Mehigan 2001, 9–17). Musil initially intended to dedicate himself, therefore, not to literature but to science: in 1903 he enrolled in further studies in psychology and philosophy at the University of Berlin. Not even the unexpected success of *The Confusions of Young Törless* could deter Musil from studying for his doctorate and from the prospect of further progress in his academic career. His interest in science may have received additional impetus from the highly promising innovations in the field of psychological science occurring at the time, many of which Musil was able to observe at close quarters during his time at Berlin University. The aim of these innovations was to furnish psychology (and other sciences) with a more reliable empirical basis. The functionalist orientation of the new science, which doubtless further heightened Musil's interest in psychology, even served as an epistemological starting point for Musil's dissertation, which, as we examined in chapter 2, was devoted to a consideration of Ernst Mach.

Mach's functionalist contributions in the field of psychology and physiology are on view in an important work *The Analysis of Sensations* (*Die Analyse der Empfindungen*), published in 1885. By the time Musil was beginning his dissertation, Mach's work was already in its fifth edition. With this work, Mach intended nothing less than a fundamental questioning of the position of metaphysics in the knowledge project, which Mach's Kantian affiliations had taught him to see as a "pre-critical" phase in the history of mentality. The move against metaphysics that Mach, following Kant, advocated also involved a move against Cartesian dualism, on the grounds that nothing could be reliably understood about the world on the basis of the difference this dualism insisted on. Instead

bloß durch Metaphorik oder Analogiebildung erzielt, sondern immer auch durch wissenschaftliche Zusammenschau der entsprechenden Spezialdiskurse [. . .]."

of such pre-critical dualism, Mach recast the knowledge project as the question of how to stabilize the agglomeration of abstract sensory data that are absorbed into or otherwise registered by the sensory complexes of human beings as they encounter and interact with their world. Since, as Kant also says, nothing final can be said about the source in the "I" that does this sensory work—Kant acknowledged it only on the basis of the successes it accomplished—the ego, in Mach's view, can only be thought of as "relatively consistent" and must remain to some extent enigmatic and unknown:

> The ostensible constancy of the ego consists above all in its *continuity*, in the fact of its slow change. The many thoughts and plans of yesterday, which are continued today, for which in wakeful perception the surrounding world continually serves as reminder (thus can the ego in dreams come across as blurry, doubled, or even be entirely absent), the small habits that are maintained consciously and involuntarily over longish periods—all these constitute the basic elements of the ego. (Mach, *Analyse*, 3)

The dissolution of the ego into de-individualized perceptive data-agglomerations represents in Mach's doctrines the first stage of the dissolution of the environment into shapes, bodies, and matter, which are for their part equally devoid of any essential individuation. In Mach's view, the fact that matter takes on the appearance of individualized forms is simply a result of practical considerations—attributable to the same apperceptive successes Kant had underscored in *The Critique of Pure Reason.* Instead of the thing-in-itself, which Mach considers to be pure delusion, Mach draws on Kant's terminology of "appearances" in order to indicate the multitude of ways in which the physical elements of the world register as sensory impressions on human senses ("impressions" is Mach's favored *terminus technicus* for appearances which he adopted from British empiricism, primarily Locke and Hume[4]). For Mach, these sensory impressions are there to begin with. This is the kernel of his difference with Descartes, who assumes that intellectual consciousness precedes the fact of physical extension. Mach proposes a starkly contrasting thesis, namely that the data of consciousness are first perceived by the senses, then transformed into sensory impressions, and finally transition into cognitive formations, which Mach calls "thought symbols" (Mach 10). Notwithstanding the operative power of these thought symbols, the leading assumption remains the fact that the world is initially made from perceptions and that entities such as bodies and egos are measures whose function is to facilitate a practical

[4] Mach 1991, 38: "That my points of departure are not essentially different from those of Hume will be clear."

orientation toward the objects of the world. This orientation is thus never permanent and remains provisional.

Despite such practical measures, which Mach thought had been erroneously turned into conditions and substances and invariable truths by pre-critically premised philosophy, Mach proposes a functionalist approach that rejects essentialist assumptions about objects and things of the world. The new focus of the analysis is that all elements of the world and all perceptions that arise in subjects are only purposive to the extent that they can be placed in relationship to other elements and perceptions—in other words, to the extent that these elements enter into functionally premised relational models (Mach 1991, 13), which, in this form, can subsequently be made available to analysis. The focus of analysis thus shifts from cognitive formations stabilized by the ego to functions located in cluster complexes beyond the ego. Mach is therefore radically opposed to traditional theories of consciousness. He cites the example of the pencil submerged in water and notes that the pencil appears one way to direct perception, another way to intellectualized understanding. Naturally we compensate for this discrepancy and distinguish between appearance and reality, but what allows us to make such a distinction? "In both cases there are facts that represent quite differently conditioned and composed arrangements (*Zusammenhänge*) of elements. The submerged pencil in relation to its environment is bent to the naked eye, but straight in terms of its palpable form and measurement" (Mach 1991, 8). Mach's challenge to metaphysics—which he believed had hindered the progress of science because it was plagued by endless debate over the body-mind split—consists, first, in strengthening the declaratory power of the senses regardless of their claims to truth, and second, in rejecting all a priori assumptions in relation to the nature of the ego. Since the ego can only maintain continuity over a limited period of time (and even then only imperfectly), views that maintain an essential form intrinsic to the ego must be surrendered for the sake of meaningful scientific analysis. In this sense, Mach asserts, the ego must be considered "unsalvageable" (Mach 1991, 19). Mach does not ascribe scientific value to an ego essence, which is inaccessible to science. Instead, he ascribes such value only to those substantive aspects of conscious experience that, by way of a non-individualized perception, present themselves to analysis in functional correlations.

## Implications Arising from Mach

Musil's intellectual outlook as a writer is thus influenced both by his early professional orientation to technical science and by his early encounter with the thought of the physicist Ernst Mach, whose psycho-physiological parallelism aimed to provide a new foundation for science at the end of the nineteenth century. While the suggestive power of Mach's ideas was clear to many writers from Musil's generation, none of them realized with greater intensity and clarity the consequences for literature that arose out of Mach's teachings.

This is apparent in *Posthumous Papers of a Living Author* (1936), a collection of short prose pieces Musil calls "pictures," "unfriendly observations," and "stories," mainly composed in the 1920s. The influence of Mach's understanding of the ego is evident, for example, in *The Blackbird*, the longest story in the collection, which takes as its starting point Mach's approach to the question of the continuity of the ego (when viewed from the standpoint of functionally understood relational networks):

> One changes . . . from tip to toe and from the hair of one's skin to the heart, but the relationship between these elements curiously remains the same and changes just as little as do the relations that each individual upholds to the various personnages (*Herren*) which he addresses, one after the other, with the word "I." (*GW* 7:548)[5]

In this story, the protagonist, "A2" (*Azwei*), wakes up in the middle of the night when he hears a blackbird imitating the song of a nightingale. As wakefulness arrives, yet also trusting the significance bound up with what he has just heard, A2 makes the momentous decision to leave his wife and "follow the nightingale." Later in the story, in a passage that relates the death of his mother, he is again woken in the night, and again it is a nightingale he hears. Half awake, half asleep, he spots a bird on the window ledge in the moonlight, identifying it not as a nightingale but as a blackbird, since, as he knows, blackbirds imitate the songs of other birds. The blackbird speaks to him in the voice of his mother. When A1, his disbelieving friend, asks him about the meaning of these events, A2 answers: ". . . it all happened just like I said; and if I knew what it meant, I wouldn't need to relate it to you. But it's as if one could hear whispering, or just a rustling, without being able to tell the difference!" (*GW* 7:562).[6]

What this story conveys is an ethic that acknowledges the key role played by the emotions in the mediation of human experience. Since it is not a sovereign ego that controls the perceptive data, but rather an ego in which the sensory data momentarily coalesce into emotional complexes before dissipating again, nothing definitive can be asserted about the value or significance of A2's stories. The communication of their value

---

[5] "Man ändert sich [. . .] vom Scheitel bis zur Sohle und von den Härchen der Haut bis ins Herz, aber das Verhältnis zueinander bliebt merkwürdigerweise das gleiche und ändert sich sowenig wie die Beziehungen, die jeder einzelne Mensch zu den verschiedenen Herren pflegt, die er der Reihe nach mit Ich anspricht."

[6] See also chapter 9, where I treat *The Blackbird* in connection with what I call Musil's "theory of correspondence."

and significance must be entrusted to a new story, "the third story, how it may end I cannot know" (*GW* 7:154). Therefore, it cannot be said with certainty that the narration has an unambiguous meaning. Instead, such a narrative would have to be augmented with further stories, whose meaning for the ego would always have to be freshly determined. But it is in this open-ended storytelling, and not in any act of final disclosure, that the possibilities of human experience are to be found.

In "Triëder," first published in 1926, a similar intention that extends beyond the usual possibilities of human experience can be identified. "Triëder" discusses the construction of an alternative perspective in relation to objects that present themselves to the ocular perspective. The effects of seeing an object through a process of bifocal mirroring[7] is described by the narrator as being characterized by a strong linear perspectivism: the binocular glasses accentuate the impression of perspectivism in a building directly opposite the viewing point of the narrator. There is even talk of the "hideous painterly myth of the disappearance of lines that rumour has further exaggerated" (*GW* 7:519).[8] The narrator does not at this point have in mind any normalization of the objects located in the field of vision via a correction of what appears to the eye. Rather, the aim is precisely the opposite: the distortions are not cleared away but preserved; of interest is precisely the disturbed view of the object. As with Mach's submerged pencil, the narrator holds to the original circumstances of seeing for the value that this might offer vision. To this extent, the "triëder" serves to found a new theory of vision:

> It [this theory] is called isolation. One usually sees things along with their environment and considers their meaning to arise from the environment they are found in. But if these things step out from this environment, they are unintelligible and terrible, just as it would probably have been on the first day after the creation of the world, before the appearances had become accustomed to each other and to us. (*GW* 7:520)[9]

---

[7] Referring to Bergmann-Schaefer's *Lehrbuch der Experimentalphysik* (1993), Claus Hoheisel describes these effects as follows: "Das Fernglas besteht im Prinzip aus einem Objektiv, welches ein Zwischenbild auf das Okular wirft. Durch letzteres wird dann das vergrößerte Zwischenbild betrachtet. Zur räumlichen Verkleinerung des Strahlenganges und zur Erzeugung eines aufrechten Bildes schaltet man ein sogenanntes Prismen-Umkehrsystem zwischen Objektiv und Okular" (Hoheisel 2009,121).

[8] In the original: "eine grauenvolle Malersage vom Verschwinden der Linien, die gerüchtweise übertrieben werde."

[9] "Sie [= die Theorie] heißt Isolierung. Man sieht Dinge immer mitsamt ihrer Umgebung an und hält sie gewohnheitsmäßig für das, was sie darin bedeuten. Treten sie aber einmal heraus, so sind sie unverständlich und schrecklich, wie es

The abiding concern of this short prose piece is thus to illuminate new possibilities of understanding at the point where objects, taken at face value, have their usual contextual significance stripped back from them. In these circumstances, one will not want to say, taking in the movement of a streetcar as it slowly wends its way toward the observer, that it follows the S-shaped twists in the tracks, but one will rather say of the box-like shape that is apprehended, following the strict perspectivism of the triëder, that it expands at certain moments, then contracts and shrinks. Similarly, one will not see pedestrians making their way in silent contemplation along the street, but rather only the clumsy forward movements of animate beings seeking to accommodate themselves to the strenuous physical demands of walking. In line with the medium of isolation, one no longer views the human being as a stable construct, upon which experience and education have done their work of organization, but rather as a kind of clothing, which according to the nature of each respective owner is either worn fittingly or not.

Musil's theory of vision thus has two aspects. First, it caters for an interruption in the circumstances of ordinary vision. In this respect it has similarities with Hume's philosophical project in the *Treatise of Human Nature* (1739), which established that the human senses are shaped by the elementary power of "custom" and "habit," and that what we are obliged to call truth, accordingly, lies beyond the socially infused patterns of vision to which we have become accustomed (Musil discussed Hume's influence on Mach in his doctoral dissertation, as we saw in the previous chapter). To put this point another way: custom provides the contextual parameters for our visual responses and hence also for our opinions and even our supposed intuitions. Musil's new theory, constructed on the basis of Mach's functionalist "antimetaphysical" line of thought, begins at the point at which such anticipatory structural models are broken down. The breakdown of the anticipations directed at the objects located in the field of vision therefore establishes the preconditions for a new optics. The technique of isolation has utility insofar as it dismantles assumptions underpinning the customary way of viewing and understanding objects. At this concluding point of "Triëder," Musil calls the claim of humans to understand objects in the visual field in humanistic terms to be "fatally cut through" (*durchschnitten*; *GW* 7:521). If this insight is given credence, then we will indeed be justified in conscripting "Triëder" in the service of a posthumanist outlook.

The second aspect of Musil's theory of vision relates to the cognitive gain that is bound up with this proto-posthumanist position. This cognitive gain results from *failing to correct* or otherwise stabilize the initial

---

der erste Tag nach der Weltschöpfung gewesen sein mag, ehe sich die Erscheinungen aneinander und an uns gewöhnt haben."

distortions presented to vision. Failing to make adjustments to perceptions, for humanistic or other purposes, will also forestall the anticipations that impinge on vision and organize the visual field through a process of "prefabrication." Musil will make the strategies of social prefabrication a focus of the later novel *The Man Without Qualities*. Stripped of their tendency toward presentiment, the data of direct vision, viewed in isolation and conceivably without distortion, will, as Musil tells us in "Triëder," amount to something "terrible, as it must have been on the first day after the creation of the world." This is the misanthropic conclusion of the prose text that led Musil to allocate it to the thematic grouping "Unfriendly Observations" in the volume *Posthumous Papers of a Living Author*.

## A New Laocoön?

The misanthropic conclusion to "Triëder" leaves the reader confronting a new problem: if we are to follow the "depragmatizing" line the text lays out for us, we must inevitably reach something like a posthumanist position in regard to our social and cultural investments. "Triëder" tells us that vision is infused with contextual distortion. It implies, though does not say outright, that such contextual distortion is culturally mediated. At any rate, the text allows us to see through to the contexts of sense that are arranged before us; these contexts of sense are variable and implicitly subject to forms of manipulation and control. The task of observing strategies of manipulation and control in turn-of-the-century Viennese society will later be absorbed into a project of arguably corrected vision in *The Man Without Qualities*.

To see without distortion, on one level, of course, is not to see at all, or else it is to see as a human might have seen on the first day after the divine creation. Mach's influence on Musil's thinking in "Triëder" in his important text *Analyse der Empfindungen* allows us to connect this ideal of seeing to the scientific enterprise. If we follow Mach, we witness an idealizing scheme at work built on a technical ideal—the "terrible" gaze onto the world when objects had no contexts by which to render them sensible. Under such circumstances, we must, strictly speaking, put the individual at a remove from the terms of vision, for the terms of human vision will always already be "pre-organized" and tend to work against the execution of the scientific enterprise. The final consequence of such a removal would be that the individual is left out of the project of science altogether—notwithstanding the plain fact that scientists themselves are also individuals. The cognitive gain from this banishment is something like unimpeded vision, a vision that promises a clear way to understand the objects that appear in the contextual field and a better way of maximizing their effective deployment in future contexts. In other words, as the individual, with his humanistic presuppositions, moves out, the scientist, with her strategies of clear and distinct seeing and the promise of advancement and scientific "progress," moves in. What sort of "repragmatizing" should follow the strenuous effort of depragmatization becomes a question for the new type of vision that aims to cash in on this cognitive gain.

This volume represents an attempt to investigate what flows from the new type of vision Musil is led to. However, as the story *The Blackbird* already suggests (and as we will investigate in greater detail in chapter 9), the new vision might make possible a new way of being in the world. This is implied in "Triëder," for example, in the image of the seemingly carefree man in his prime, whose gait is illustrative of the autopoietic traits of animal life. Humans turn out to be not, according to the old humanism, creatures of a benevolent God, but rather "a human system [. . .] only concerned to preserve itself, without anything left over for others" (*GW* 7:522).[10] The ironic undertone resonating in this passage indicates that the new optics requires critical reflection if it is to become useful for human beings. For Musil, the new optics leads to results which at the outset are not clearly discernible as good: on one hand it contributes, as the narrator reports, "to the process of understanding the individual human being"—and perhaps, therefore, a new Laocoön. But on the other hand it also contributes "to a growing lack of understanding for humanity" (*GW* 7:522) when seen as a whole. Whether the functionalist benefit of insight in the former case outweighs the civilizing loss of knowledge attaching to the latter is one of the questions put to discussion throughout Musil's oeuvre.[11]

---

[10] "ein menschliches System [. . .], das nur darauf bedacht war, sich selbst zu behaupten, und gar nichts für andere übrighatte!"

[11] Kordula Glander is of the view that Musil's later texts, namely those which were published after *Unions* (1911), can be seen as "an increasing emancipation of the literary" with respect to science (Glander 2005, 135). My view, by contrast, is that the tension between literature and science is never completely resolved in Musil's works.

# 4: The Essays

ROBERT MUSIL'S ESSAYS fall into two distinctive phases of his life and work: the period before the outbreak of the First World War and the period after the war. That Musil made use of the medium of the essay at all can be attributed to his literary program that had stalled before the war: he wanted to clarify his literary intentions and protect his experimental stories, published under the title of *Unions* in 1911, against crude misunderstandings. We can thus read into Musil's turn to the essay form an educational intention, at least to begin with, that sought both to highlight points of difference in relation to the work of his contemporaries—one thinks particularly of Musil's discussions of Spengler's thought between 1918 and 1921—and also to call attention to related or sympathetic literary undertakings, such as those of Robert Müller.[1]

Nevertheless, if one views the essays in toto, one sees that their educative intention increasingly loses its sharp edge over time. The tenor of Musil's writing in these essays changes, I believe, because Musil begins to see the essay not merely as a vehicle to promote his literary interests, but increasingly as a valuable literary form in its own right. Moreover, if the *Unions* project had run aground on the suspicion that imaginative writing could not—and perhaps should not—be used for an overriding conceptual or "scientific" purpose, then Musil, as a result of the poor reception given to *Unions*, was obliged to consider alternative literary forms that might do a better job in helping to fulfill such a purpose. The essay form showed promise in this regard. On the face of it, the essay is a "cool" medium that sets out to promote dispassionate reflection on the topics that make up its subject matter—not infrequently matters of current social or political concern. If the essay could now be put to work alongside literature, where topics and themes are broached from the perspective of the emotions—an avowedly "hot" medium—then Musil would be much closer to realizing his ambition to unify conceptual and emotional reflection. This unification would be built on assumptions that only the combination of concept and emotion would be able to deliver the kind of insights necessary for the development of ethical understanding. If we take Musil's reflections on the work of Franz Blei as an example (bearing in mind Blei's own role in stimulating Musil's *Unions* project, as I

[1] Musil's diaries contain several references to Müller, particularly in the period 1918–21.

will discuss in chapter 7), we see the kind of change in the deployment of the essay just mentioned. Instead of the approving, but rather cool and distanced, manner of discussion we encounter in the discussions of Blei's work in an essay that appeared in 1918, an essay thirteen years later reveals a more studied interest in Blei's intellectual outlook that links his writing to the philosophy of the Enlightenment and expressly acknowledges the role accorded to the "formation of ideas" (*GW* 8:1201).

In the 1918 essay on Blei Musil had registered the insight that the goal of an essayist is the "articulation of feeling by the mind, the capacity to turn the mind away from the meaningless tasks of knowledge toward the task of feeling" (*GW* 8:1014). But we do not discern in these remarks an interest in the form of the essay itself. Yet when Musil comes to the defence of writers in his 1931 essay "Authors and Literature: Marginalia" ("Literat und Literatur: Randbemerkungen dazu"), describing them as more than "just writers," it is clear that he has come to understand the task of the writer in broader terms, now emphasizing that the author wears many hats—those of the "dramatist, historian, critic, essayist, and so on" (*GW* 8:1204). In line with this expanded conception of the task of the writer evident from the mid-1920s on, Musil characterizes his own ambition in literature in an interview with Oskar Maurus Fontana in 1926 as an attempt to "deal with the world intellectually" (*Beiträge zur geistigen Bewältigung der Welt geben*).[2]

The direct expression of this intellectual program of "dealing with the world" is *The Man Without Qualities*, a novel whose origins reach back to the turn of the century, but whose major themes were only worked out in the period following the end of the First World War when Musil returned to civilian life after serving on the Italian front. The major intellectual achievement that this novel represents would hardly have been possible without the deep reflective tissue from which the novel is made. This tissue—a mix of plot and argument, vivid image-making and dispassionate intellectual inquiry—is possible not only because of the ideas that are patterned into the structure of the novel at a deep level but also because of the deployment of the essayistic mode itself, which precisely enables such patterning to occur.[3] With this later use of the essay and the ideas they transport in mind, Musil's essays, I suggest, are to be understood less as a series of interesting but largely peripheral intellectual excursions sitting alongside Musil's more important literary production, and more as an absolutely central component of his ambition to broaden the range of

[2] "Was arbeiten Sie? Gespräch mit Robert Musil [30. April 1926]" (*GW* 7:942).

[3] In the interview with Fontana Musil concedes that the deployment of the essayistic medium gave rise to certain "fears" about how it would meld with the literary aspects of the major novel. But these fears only indicate the importance of the medium for the project Musil was intent on pursuing (cf. *GW* 7:941).

literature's conceptual and intellectual resources in the interests of a wider intellectual program.

## "Sketch of the Writer's Knowledge"

Among Robert Musil's vast output of essays in the period to 1931, the essay from 1918 entitled "Sketch of the Writer's Knowledge" ("Skizze der Erkenntnis des Dichters") commands an especially prominent position. This essay is noteworthy for the attempt it makes to determine and clarify the kind of knowledge claim that literature is capable of issuing. This attempt at clarification involves two steps. In the first step, Musil denies that literature is cut off from the region of the conceptual in point of principle. If the basic assumption of the conceptual domain is that ideas stand apart from the medium in which they are presented, Musil's claim here is that literature can make a serious contribution to the presentation of ideas *precisely because these ideas cannot wholly be separated from the language in which they gain expression*. Musil's approach to the conceptual thus entails the "Whorfian" claim that the medium of reflection promoting the appearance of ideas will have something to do with the ideas themselves and the type of insight derived from them. This claim means that concepts and the texts from which they are derived cannot be imagined to run along separate tracks.

Before we come to the second step in Musil's argument, it is worth consulting similar reflections on the question of the relation of concepts and texts put forward by Stephen Toulmin. As we know at least since Kuhn's *Structure of Scientific Revolutions*, the question of this relation and how it should be framed is metaphilosophical. Toulmin puts the question in these terms: "Are the meanings philosophers set out to analyze embodied in isolated verbal propositions, or is language intelligible only as having a meaning within the larger framework of actions and institutions?" (Toulmin 2001, 11). When we review Musil's essay "Sketch of the Writer's Knowledge," we see that Musil takes up a position that rejects the former and endorses the latter. Meaning—scientific meaning—is not to be imagined as divorced or "desituated" from the wider context from which it emerges and which it addresses. Yet through a historical process that started in the early seventeenth century, precisely such a divorce, Toulmin argues, began to occur as "the prestige of mathematical proofs led philosophers to disown non-formal kinds of human argumentation" (14). To collapse a long story into a short thesis, Toulmin identifies a separation between a "rationality" that promoted and deepened interest in formal proofs and a sense of ordinary "reasonableness" that increasingly lost prestige as a form of "human argumentation": "Beside the rationality of astronomy and geometry [in the seventeenth century], the reasonableness of narratives came to seem a soft-centered notion, lacking a solid basis in philosophical theory, let alone substantive scientific support" (15).

When we consider Toulmin's argument that the debate about resetting the terms of this imbalance between "rationality" and "reasonableness"—both of which can be expressed in German by the term *Vernunft* ("reason")—was not properly settled until the twentieth century, in the period from 1920 to 1990, we already see Musil's prescience in this regard. For Musil's point in his 1918 essay is precisely to remark upon the historical emergence of "rationality" (in Toulmin's sense) and to note its formal quality as a one-sided application of rule-following. This is then the second step accomplished in Musil's essay. Since, as just noted, both "rationality" and "reasonableness" are encompassed within the semantic reach of the German word for "reason," Musil comes up with a new word—the neologism "ratioïd"—to describe the characteristics of the type of thinking that has now become established in his age and for which science has reserved special prestige:

> This ratioïd domain—crudely delineated—embraces everything that can be systematized scientifically into summary laws and rules, above all physical nature; it includes the moral only in certain exceptional cases of success. It is characterized by a certain monotonous factuality, by the prevalence of repetition, by a relative independence of each fact from other facts, with the result that facts usually conform to groups of laws, rules, and concepts that have been established in advance, in whatever order in which they were discovered. Most notable in this regard is that the facts in this domain can be clearly described and communicated. (*GW* 8:1026–27)[4]

The "ratioïd" domain is governed by the logic of scientific rules, and can be contrasted with the "non-ratioïd" domain, which tends to take the form of "the sovereignty of exceptions over rules" ("Herrschaft der Ausnahmen über die Regel" [*GW* 8:1028]). The distinction Musil has in mind does not just concern the results that are derived from these differently premised domains, where, for example, the productive yield from one domain might be "science" and forms of "art" might be consistent

[4] "Dieses ratioïde Gebiet umfasst—roh umgrenzt—alles wissenschaftlich Systematisierbare, in Gesetze und Regeln zusammenfaßbare, vor allem also die physische Natur; die moralische aber nur in wenigen Ausnahmsfällen des Gelingens. Es ist gekennzeichnet durch eine gewisse Monotonie der Tatsachen, durch das Vorwiegen der Wiederholung, durch eine relative Unabhängigkeit der Tatsachen voneinander, sodaß sie sich auch in schon früher ausgebildeten Gruppen von Gesetzen, Regeln und Begriffen gewöhnlich einfügen, in welcher Reihenfolge immer sie entdeckt worden seien. Vor allen Dingen aber schon dadurch, daß sich die Tatsachen auf diesem Gebiet eindeutig beschreiben und vermitteln lassen." (*GW* 8:1026–27)

with the other. It is accordingly important to assess what happens to thinking when the inputs governing it are varied in particular ways. Thought in the ratioïd domain, for example, is rule-bound and sequential. One might say today that it is achieved by the programming of logical flow sequences typical of the digital machine. Musil's speculations about the ratioïd, indeed, seem to anticipate digital computation—the question of whether in a given logical sequence an input is consistent with the rule it follows or at variance from it (at which point a digression into a new rule chain might apply and be laid down). At this point, Musil does not find the ratioïd domain wanting or deficient: he neither affirms nor invalidates it, implying that the domain will command utility in the special (mechanical or digital) contexts for which it has been fashioned.

Nevertheless, Musil does underscore what this kind of thought cannot achieve, or, at any rate, does not normally achieve. Musil points out, for instance, that the ratioïd domain will not usually be able to cross-connect with moral thought—the moral occurs in this domain, he says, "only in exceptional cases of success." This is a crucial observation that throws light on Musil's conception of the nature of ethics. The basic point is this: we will not be able to expect ethics to be consistent with a principle of rule-following. This means that the appearance of the moral will not be governed by any formal procedures construed in advance that might be designed to contain, express, or capture its essential form. In Musil's eyes, the domain in which the moral appears, the domain of ethical thought in general, is the non-ratioïd domain, a mixed domain where thought encounters feeling and in which exceptions trump rules and sometimes make them superfluous. By this argument Musil moves against the entire tradition of Kantian thought, which attempts to think of the ethical in terms of the strictures provided by a "super rule," which is to say, the application of a principle to thought to which such thought, in order to be ethical, must conform. Musil therefore effectively denies that Kant's so-called "categorical imperative" provides a binding principle for ethical thought in general.[5]

This does not mean, Musil argues, that the non-ratioïd domain gives up all connection with knowledge and the realm of ideas. On the contrary: this domain is described not just as "a domain of reactivity of the individual against the world and other individuals"—a reactivity issuing from the "unsalvageable isolation of the self (*Ich*) in the world and between human beings" (*GW* 8:1026)—but rather also as "the domain of values and value judgments, the domain of ethical and aesthetic relations, the domain of

[5] A note from his diary, volume 19 (1919–21) indicates Musil's familiarity around this time with Kant's moral theory and its importance in a national context: "The moral comes from Kant, floating in national sentiment" (in the original German: "Das Mor[alische] kommt von Kant, suspendiert in Nationalgefühl" (*TB* 1:535).

the idea" (*GW* 8:1028). The idea manifests itself in ways fundamentally dependent "on the manner of its application and on the person [applying it]," and has only a "contingently determined meaning that disappears when it is removed from its accompanying circumstances" (ibid.). In the essay "Sketch of the Writer's Knowledge," then, Musil defends a type of writing by which new modalities of thinking and feeling come into view. In this domain, reasoning (as "reasonable" thought in Toulmin's sense) and moral thought are brought together. Musil subsumes both under the guiding principle of the "idea," a manifestation of thought directed toward (ethical) understanding that is not mired in rationalistic conceptualizations. The idea comes alive through the creative power of an agency whose task in the non-ratioïd domain—as Musil then explains—is working out the role and significance of the fact of singular experience: "Facts in this domain are not preeminent, laws are sieves, events do not repeat themselves but are unrestrictedly variable and individual" (*GW* 8:1028).

The implications flowing from this analysis of the ratioïd and the non-ratioïd cannot be underestimated. For one thing, the confidence that mathematical thought, as a species of thinking in the ratioïd domain, can provide ethical orientation for the human being—one of the sources of Törless's confusions—must be formally abandoned. When we call to mind *The Confusions of Young Törless*, Musil's first novel, we note that it was the mathematics teacher who recommended that Törless read Kant to dispel his confusions. We are now in a position to say from a different position why Törless gets nothing of value from Kant: the ethical, in Musil's conception of this domain, is not dominantly determined by rules and rule-following. Musil—in line with the thought of Nietzsche, as will be discussed in chapter 6—moves ethics into a post-Kantian orbit in which it will be attached to thinking that expressly eschews rule-adherence. Ethics, in this understanding, will have no predictive quality, will not be generated from the province of principles and precepts (though might of course consult them), and at most will have certain notable features that illustrate it—the kind of characteristic features that the novellas of *Unions* sought to make evident.

If the need to identify the features of the ethical leads into the non-ratioïd domain, it also becomes clear why Musil will thematize the fact of separation of the "scientific" from "ordinary" understanding at the start of the novel *The Man Without Qualities*. This novel famously begins with a technical account of the weather, using meteorological data in relation to a particular day in August, and contrasts this account with the perspective about the weather that rises in ordinary understanding. For if the region of the ethical does not in principle overlap with the domain of the ratioïd or is not (any longer) covered by it, then the human being who seeks an ethical comportment in life will be obliged to confront a separation between the technical domain of the ratioïd and the non-technical domain of the non-ratioïd to which ethics must now default—all the more so when that human being, the protagonist of Musil's novel, is a mathematician. The contrast between these domains is sharpened by the fact that the type of thinking hitherto held to govern the ethical, Kantian ethics, is drawn from precisely the kind of fidelity to rules—the "ultimate" technical rule of the categorical imperative—that, to Musil's way of thinking, no longer provides anything useful for true ethical understanding or action. The question Musil therefore asks in the novel, in view of this situation in which ethics in a post-Kantian era is no longer understood to be governed by the mechanism of rule-adherence, is a return to prominence of an ancient question—the question of how a human being is to live. This question, in turn, becomes the main point of focus of the major novel. In his interview with Fontana Musil puts the question that emerges from this background of intellectual change in the following way: "How is an intellectual person to behave toward reality?" (*GW* 7:940)

## "Intellect and Experience" and "Steps toward a New Aesthetics"

In several important essays written in the early 1920s Musil outlines the nature and extent of the "non-ratioïd" domain. Of particular importance are the essays "Intellect and Experience: Notes for Readers Who Have Escaped the Decline of the West" (1921; Geist und Erfahrung: Anmerkungen für Leser, welche dem Untergang des Abendlandes entronnen sind) and "Helpless Europe or Getting Further and Further Off Track" (1922; Das hilflose Europa oder Reise vom Hundertsten ins Tausendste). In these essays, defining how far literature does justice to ideas and can meaningfully deploy them becomes an abiding concern. For example, in "Intellect and Experience," Musil holds Oswald Spengler's rejection of scientific thought in *The Decline of the West* (*Der Untergang des Abendlandes*), the first part of which had appeared in 1918, to be exaggerated. Musil summarizes Spengler's argument in this work in the following terms: "Reality doesn't exist. Nature is a function of culture. Cultures are the last known available reality" (*GW* 8:1045).[6] Against such skeptical relativism, which ultimately seeks to do away with the scientific enterprise, Musil offers a limited empiricism, one that assumes "a common regularity" behind objects. This is a regularity "that lies beyond the subject and is thus an experience that would be capable of extension and refinement, of the possibility of being known, of a version of reality, progress, advancement, in short: precisely the kind of mixture of subjective

[6] "Es gebe keine Wirklichkeit. Natur sei eine Funktion der Kultur. Kulturen seien die letzte uns erreichbare Wirklichkeit" (*GW* 8:1045).

and objective knowledge whose separation constitutes the arduous work of evaluation found in a theory of knowledge" (*GW* 8:1045, emphasis in original).[7]

Musil does not wish to throw out the "arduous work of evaluation" to which formal (and especially scientific) thought is directed. Nor does he want to ratify the denial of every empiricism and every conceptual orientation for thought. At the same time, Musil is a long way from following a facile empiricism that is calculated to put the intellect in chains. The clarification in regard to empiricism Musil undertakes in this essay rather consists both of making concessions to rationalistic positions, to the extent that these are consistent with a notion of experience necessary for the experimental process, as well as critiquing precisely those positions that reveal a "narrowly scientific spirit, intellectualism, rationalism, and so on" (*GW* 8:1049). In the course of this undertaking Spengler proves to be useful, for he illustrates the shortcomings attaching to a dogmatic rejection of scientific reasoning as well as the virtues of an ethical discussion rooted in the type of assumptions typical of the non-ratioïd domain. Despite sometimes appearing to give Spengler short shrift, then, Musil accepts that there are some advantages to Spengler's approach, principally in relation to the need to find or form a new ethical "relationship to the world and humanity" in general (*GW* 8:1054).

In the essay "Steps toward a New Aesthetics: Remarks on a Film Dramaturgy" (1925) (Ansätze zu neuer Ästhetik: Bemerkungen über eine Dramaturgie des Films), Musil uses the example of film to reflect further on the skepticism that science practices in its methodology and the certain skepticism about and *against* science that this skeptical method incites. Once more Musil adopts a historical and anthropological viewpoint in order to arrive at an intellectual assessment of what he calls "the normal condition of our relations to the world, to other people and to our own ego" (*GW* 8:1143). In Musil's view, it is precisely "the acuity of our intellect"—which is to say, the skepticism that science codifies as part of its method—that has seen humanity develop to the relatively advanced position it is found at today: "Rulers of an earth on which, among monstrous beings, we ourselves were once nothing" (*GW* 8:1143). But it is on the strength of this same intellectual acuity—a mental capacity Musil characterizes as a "measuring, calculating, sensing thing, a positive, causal, mechanical way of thinking"—that has led to some of the problems now suffered by humanity, problems that threaten to end up in a deep distrust of, and bitter

[7] "Das außerhalb der Subjekte liegt, also eine Erfahrung, die der Erweiterung und der Verfeinerung fähig sein könnte, die Möglichkeit eines Erkennens, irgendeine Fassung von Wahrheit, des Fortschritts, Aufstiegs, kurz gerade jene *Mischung* subjektiver und objektiver Erkenntnisfaktoren, deren Trennung die mühselige Sortierarbeit der Erkenntnistheorie ausmacht" (*GW* 8:1045).

skepticism toward, all aspects of the external world and the attempt to have knowledge of it. Musil's preliminary sense of how to navigate among the shoals of modern life in view of this skepticism against science is this: one must ratify the results of mechanical-causal thought and its methodological "mistrust with ancient roots," on the one hand, but we must wean ourselves off its dominance on the other. We must therefore come to a mature understanding of both the limits and the virtues of science. Thus instead of demanding an increase in the deployment of "mechanical-causal thought" and thus "more science," Musil advocates a more balanced position by referring to the significance of the historically "no less proveable" (*GW* 8:1144) other side of the human intellect, which he calls "the condition of love . . ., of goodness, of distance from the world, of contemplation, inner reflection, approach to the divine, of idealism, of refuge (*Einkehr*)" (ibid.). However refined mechanical-causal thought and calculation may have become over the course of human history, then, the qualities and advantages of this other side of human experience appear to Musil to have fallen to a certain extent by the wayside. This alternative dimension to our thought and feeling, which Musil connects in this essay for the first time with the term "the Other Condition" (*der andere Zustand*), a descriptor later to play such a prominent role in the major novel, has, as Musil now avers, largely remained at the point of development it commanded a thousand years ago. There has been little progress achieved in terms of understanding its potential for use since this time. For this reason it is now urgent that we revise our approach to this other dimension of our being and thereby transform our entire intellectual outlook.

By the same token, Musil does not see his age as suffering in general from too much scientific thought, though he does concede that there are problems issuing from its one-sided deployment. The tendency Musil criticizes lies with the practical prescriptiveness of the scientific in the cultural domain—what he calls in the essay "Steps toward a New Aesthetics" its "orientation toward formulaic thinking" (*GW* 8:1146). In the second part of the first book of the major novel Musil will treat the complex issue of formulaic thinking, which has resulted from an over-reliance on the domain of the ratioïd, under the heading of "Semblances Occur" (*Seinesgleichen geschieht*). This term is meant as a gloss on the general difficulties besetting the thought of his age. In the essay "Steps toward a New Aesthetics," on the other hand, he closes off the path that would lead in the opposite direction toward something "dark" and opaque, merely a type of mystical experience. Instead of mysticism, Musil offers the virtues of an artistic understanding that aims at a new clarity of understanding. The *non plus ultra* of this type of understanding is the individual and the life experience that is given only to individuals. Musil calls this experience the "never wholly repeatable, not determinable, . . . even anarchically lived experience

(*Erlebnis*)" (*GW* 8:1151) of the individual,[8] an elevated condition of being linked to "another totality," "another dimension of experience." This other dimension only becomes visible via internal pathways through the self, where the world "is not experienced as a connection of reified relations but as a succession of ego-related experiences. The vector, the path of direction . . . has turned around and faces toward the inside" (ibid.).[9]

## "Authors and Literature: Marginalia"

The essays written up to the mid-1920s can be read as an attempt to outline the new mission of art later exemplified in the major novel. In "Steps toward a New Aesthetics" Musil finds words such as "excitation" and "feeling" to highlight the momentary "improvisation" or variation of the "normal condition" of life animating this conception of art, notwithstanding that such a normal condition, the usual conduct of life, never recedes from attention in the novel. The task of the ethical condition this new artistic conception Musil aims at is an emotional enhancement of life that enriches and redirects the usual condition of life.

The last step in the expounding of this new understanding of art is taken in Musil's last essay "Authors and Literature: Marginalia" (1931; Literat und Literatur: Randbemerkungen dazu). In this essay Musil distinguishes literature from scientific forms of communication in a way that might lead to the assumption that a dualism of the ratioïd and the non-ratioïd—a dualism of substance or substantial forms—is ultimately being aimed at. Yet such a dualism is quite far from Musil's true intentions. A dualism of substance, for example, would bring about the conclusion that one side of the dichotomy—say the ratioïd side—would ultimately have to be rejected as not being conducive to the inner development of the human being. In "Authors and Literature" Musil counters such a view by redefining the relationship between words and concepts. Words are not straightforward vessels for concepts, Musil argues, which cannot be retrieved as if it were solely a matter of consulting a particular word and unpacking its contents. Rather, a word is "merely the seal on a pack of loose representations" (*das Siegel auf einem lockeren Pack von Vorstellungen*, *GW* 8:1212–13). When we breach this seal, we find, Musil suggests, a plethora of significations but no single determinate meaning. We must accordingly revise what we hold concepts to amount to, as well as revise our conception of the work that needs to be done in order to

[8] "Niemals gänzlich wiederholbares, nicht fixierbares, individuelles, ja anarchisches Erlebnis" (*GW* 8:1151).

[9] "Die Welt wird nicht als ein Zusammenhang dinglicher Beziehungen erlebt, sondern als eine Folge ichhafter Erlebnisse. Der Vektor, der Richtungsanzeiger . . . hat sich umgekehrt und ist nach innen gerichtet" (*GW* 8:1151).

make conceptual meaning evident in the contexts in which concepts are encountered or made. Musil then likens the work of the imagination that is needed to accomplish this work of building meaning through concepts to a bridge that "swings . . . the further one takes steps across it" (*GW* 8:1213).

One might conclude from such an image that the process of constructing conceptual meaning is quite arbitrary. Yet Musil does not reach such a conclusion. Although language is "ambiguous" and in some ways indeterminate, our use of this language need not be. Musil defends a relational conception of language that underscores that meaning emerges, first, from the larger patterns of human life (cf. Toulmin 11), and secondly, from the precision with which we observe the patterns and formations language reveals to us when we use it. The task at hand is to critique these relations. Musil contends that an internally consistent sense of relation between such determinations will in fact typically arise. These determinations do not perforce lose touch with reason. The relationship between art and rationality, as Musil emphasizes in this essay, is thus not intrinsically "one of opposition" (*GW* 8:1221). Once again, Musil applies a historical and anthropological lens in order to suggest how the complex metaphilosophical relationship between art and rationality, or, as we might prefer to say today, between the discourses of artistic and scientific thinking typical of two broad "cultures" of understanding, has come about:

> For within living memory in the field of individual psychology as a result of diverse influences, the received, highly rationalistic conception of the human mind that slavishly follows a pattern of logical thought has been replaced by a more decentralized picture according to which most of an individual's decisions are not rationally or purposively or hardly even consciously taken but occur as a result of reactions of, as it were, closed sub-parts or, as they are also called, "actional complexes" (*Leistungskomplexe*), which respond to specific circumstances. [. . .] That should not be understood as a divorce [of these parts] from the rational mind, but, on the contrary, underscores what is meant by consciousness, reason, the person, and so on [in the first place]. (*GW* 8:1221)[10]

[10] "Denn im Verlauf des letzten Menschenalters ist in der Psychologie des Ich, von verschiedenerlei Einflüssen bedingt, an die Stelle des überlieferten, sehr rationalistischen und unwillkürlich dem logischen Denken nachgebildeten Seelenschemas [. . .] allmählich ein Bild der Dezentralisation getreten, wonach jedermann weitaus die meisten seiner Entscheidungen nicht rational, nicht zweckbewusst, ja überhaupt kaum bewußt vollzieht, sondern durch Reaktionen von sozusagen geschlossenen Teilen, 'Leistungskomplexen,' wie man sie auch genannt hat, die auf bestimmte Umstände 'ansprechen' [. . .]. Das ist nicht im Sinn einer 'Enthauptung' zu verstehen, im Gegenteil, die Bedeutung des Bewußtseins, der Vernunft, der Person usw. wird dadurch gekräftigt [. . .]." (*GW* 8:1221)

This "decentralization" of the spirit, which has resulted from historical processes, must, in Musil's view, be accompanied by an art of equally decentralized proportions. The aim of this decentralization would be to address the bias toward rationality in Western culture, because, as modern psychology now tells us, those who live in the industrialized West do not decide about important life issues primarily on "rational" or purposive grounds, as is widely assumed. Yet that does not mean that these denizens of the developed societies of the West are prone to fall victim to the opposite—irrationality, whose philosophical expression is relativism. Rather, for Musil, the role of thought and also of the "non-ratioïd" lies in a project that seeks to redefine the relation beween reason and irrationality, emotion and logic in profound ways. This redefinition does not turn on the conception of the human being as a kind of "thought-machine" controlled by the superior forces of reason, but seeks to respond to that being in the full flux and flush of life, where the things that impinge upon that being are a mix of both consciously and unconsciously sensed objects and forces in situated settings. Musil understands the processing of these diverse and variously received and comprehended objects and forces as an ethical process that can only be worked through at the level of the "actional complexes" that impinge upon the ego. Manifesting such complexes becomes the task of a new—itself programmatically open and procedurally "decentralized"—art. The task of making proper determinations on the basis of an understanding of these forces within the self, then, falls to an art of the non-ratioïd. In the later major novel, where this awkward neologism is no longer used, Musil will call the domain of significance in which art seeks to make the sort of determinations that are better fitted for the task of living in modern life "the Other Condition."

Musil's essays can therefore be seen as an important precursor of the major novel *The Man Without Qualities.* Although this novel has a long gestation going back three decades, its decisive shape can be traced to impulses that only became available as intellectual material to the author in the period after the First World War. In these essays the author incubates a new understanding of art, which seeks to merge the plane of ideas with the conventions of narrative literature. The metaphilosophical framework for the novel was developed in key essays written during this time, which is to say, in the period between 1918 and 1931.

In working out the need for a new conception of art in modern times, Musil makes several limited forays into the history of ideas. A crucial discovery in this respect is the finding that a new type of hyperrational thought has emerged that has implications for thought in general. Musil coins the term "ratioïd" to describe this hyperrational thought. The appearance of the ratioïd, a new event in intellectual history, makes urgent the kind of project that Musil puts forward in the later major novel. This project is constructed around the urgent need to achieve clarification of the way human beings think, feel, and act. Musil then comes to a second insight that has enormous significance for the conduct of individual life. This is the finding that the ratioïd domain typically excludes the ethical. If ethics is not reliably to be found within such a domain—one now dominated by an order of consistency arising from the application of hyperrational procedure to the objects it construes as important—then the question arises: where it is to be found? As Musil reasons, ethical determinations can only be discovered—or are only made—in the domain of the "non-ratioïd." This domain is typically the domain of singularity. The dominant aspect of this domain is the fact of singular human experience—the fact that, for a single human life, nothing repeats itself except by accident. The domain of the non-ratioïd, accordingly, is the domain in which individual life realizes its individuality in contingent ways. It is in this domain that ethics is to be found, or, perhaps we might better say, "invented." Because in Musil's description of the non-ratioïd the overriding task will be a new approach to the ethical—one that expressly turns away from Kantian conceptions of a categorical moral law held to predetermine what ethical thought must be taken to be. Musil's conception of the ethical breaks with Kant's moral law in point of principle.

The distinction between the contrasting ways of thinking that underpin Musil's conception of the ethical—the "ratioïd" and the "non-ratioïd"—is introduced by Musil in the important essay "Sketch of the Writer's Knowledge" (1918). Musil appears to have understood the new art he advocated as a medium of reflection of equivalent intellectual stature to the procedurally rule-bound work of the ratioïd domain. On the one hand, this is possible in a basic way, because the non-ratioïd domain retains a connection to "older" forms of reason and—we might say with Toulmin—with the historical terms of "reasonableness." On the other hand, as the discussion of *Unions* in chapter 7 will show, art is not cut off from the realm of ideas but merely develops ideas in a different way from that typically pursued by science. In a project that was to become the vast intellectual endeavor of the major novel, Musil appears to have been led by the conviction that the ideas worked up in the non-ratioïd domain of art stand on their own ground and can be defended as a legitimate form of knowledge. The claim of these ideas to constitute a form of knowledge rests in part with the precision that is applied in the process of making them evident. The result of insights developed in this domain of thought and feeling would constitute, as Musil conceived this project and brought it to bear on the major novel, a genuine contribution to "dealing with the world intellectually," and genuine progress in response to the ethical question of the right way to live in modern times.

To conclude this picture of the type of thinking that Musil defended as a form of knowledge in its own right, it is helpful to consider passages

Musil had in mind when he wrote the foreword to the last work published in his lifetime, the *Posthumous Papers of a Living Author*. In the last unpublished variant of the foreword written in 1935, two years after the second release of *The Man Without Qualities*, Musil defended the art he advocated in the following terms:

> Poetry provides sense-images [*Sinnbilder*]. It is sense-making [*Sinngebung*]. It is interpretation of life [*Ausdeutung des Lebens*]. Reality is its material. (But it also provides models. And it makes partial suggestions.)
>
> Two questions are bound up with this: a) What is meaning? b) Does art do nothing beyond this?
>
> In relation to a): meaningful grasping [*sinnvolles Erfassen*] is something different from sober comprehension. It involves not only an order of understanding, but above all an order of feeling. Sense-making is in all cases [a process of] supplying the terms of inner life [*innere Lebensgebung*]. (*GW* 7:970)

The focus of the second section of this book will accordingly be Musil's attempts to provide clues, or, as he puts it, "partial suggestions," in relation to this process of inner sense-making.

# Part II.

# Aesthetics and Ethics in the Context of the Two Cultures

# 5: Modernity's Crisis

In 1977, when postmodernism was being received into American universities, Gerald Graff wrote an article in the *American Scholar* under the title "Fear and Trembling at Yale." "It is possible," he wrote, "that the true source of the malaise of current criticism is not so much epistemological or ontological as institutional, that it is a reflection of the confusion of the literature department. . . ." He went on to say: "Perhaps we have not really lost our capacity to interpret literary texts with relative correctness, but simply lost interest in this project. If so, it is not the metaphysical groundlessness of criticism that is at issue, but its pointlessness, at a moment when the humanists do not seem to know what they are trying to do" (Graff 1977, 476). Of course, the groundlessness of metaphysics, formally conceded by Heidegger,[1] has been familiar to us for a long time. What we knew less in the 1970s, or did not wish to acknowledge, was that art and literary criticism are groundless for the same reasons.[2] For a time, Gadamer's hermeneutics and Adorno's "material aesthetic" had us hoping for a final-hour reprieve. When neither strategy, nor any other like it, could legitimize criticism and thereby prevent the fall of the sovereign critic, the concession that criticism has no provable ground was less remarkable than the fact that, virtually overnight, it was universally proclaimed. It was French scholars, reading Heidegger and Nietzsche—German philosophers widely disdained in the immediate postwar period—who first drew attention to the fall of criticism by announcing the arrival of anti-exegetical practices such as "deconstruction" and "discourse analysis" in view of "the death of the author" and the "logocentric," if not "phallocentric," predisposition of texts. What they conveyed to us in sometimes brilliant, sometimes verbosely

[1] Heidegger's analysis of the groundlessness of the concept of reason is contained in a number of his writings, among them his *Kant and the Problem of Metaphysics* (1962). There he states (221): "in pursuing his laying of the foundation, Kant undermined the base [*Boden*] on which in the beginning he set his *Critique*. The concept of pure reason and the unity of a pure sensible reason become problems. Kant's profound study of the subjectivity of the subject, 'the subjective deduction,' leads us into obscurity."

[2] Of "judgment," that is, criticism in an extended sense, Kant tells us in the *Critique of Pure Reason* (against views that profess to prove God's existence): "The unconditioned necessity of judgments, however, is not an absolute necessity of things" (Kant 1990, 2:530: A 593/B 621).

impenetrable prose, was that attempts since the Enlightenment to make reason the basis for our sense of ourselves and our institutions are part of a control reflex issuing from reason itself. Reason, accordingly, cannot be used to legitimate what derives from reason, which ends up being almost everything that we take for granted in modernity, including the critical faculty. The recognition that we cannot use reason to legitimate our institutions, nor any other aesthetic value that rides with them, was the public event we now call postmodernism. To the extent that we share an understanding of what flows from the crisis of reason as a justificatory or metaphysical concept, we have all made the postmodern turn. Whether we should also share a sense of collective "ennui" as a result of this turn to the postmodern, that is, whether groundlessness automatically leads to pointlessness, is, however, quite another matter.[3]

In this chapter I address the groundlessness of criticism in relation to Musil's vast unfinished project-novel *The Man Without Qualities.* Musil published the first part of his major novel at the end of 1930; Heidegger's treatise—the signature text of European existentialism—appeared in 1927. What links the writer-philosopher Heidegger and the philosopher-writer Musil is that both authors in these works anticipated every aspect of the moment of groundlessness of modern being[4] and went through it, albeit in different ways. Since the institution of criticism has now caught up with this moment and indeed proclaimed its crisis, as Graff's comments reveal, it is instructive to consider how critics have viewed Musil and what implications they have drawn for the conduct of criticism in view of, and after, the turn to the postmodern, which is to say, in view of the end of the age of criticism. To this end, in the second part of this chapter I consider two recent critical approaches that have radicalized the moment of groundlessness in Musil's writing in notable ways: Albert Kümmel's study *Das MoE-Programm* of 2001,[5] subtitled "A Study of Intellectual Organization" (Eine Studie über geistige Organisation), and Stefan Jonsson's monograph *Subject without Nation*, published in the year 2000, which is subtitled "Robert Musil and the History of Modern Identity." Before I engage with these readings of Musil, and preliminary to doing so, let me return to the postmodern moment and what, on certain accounts, it holds within it.

---

[3] The recent interest in boredom as a philosophico-literary problem registers the significance of the issue of groundlessness in criticism. Cf., e.g., Kuhn 1976, Mandelkow 1999.

[4] Heidegger restated the problem of the groundlessness of being in his late work *Die Grundbegriffe der Metaphysik. Welt–Endlichkeit–Einsamkeit* (1983/92), in which groundlessness and boredom ("Langeweile") are related moments of modern being: "Do things ultimately stand in such a way with us that a profound boredom draws back and forth like a silent fog in the abysses of Dasein?" (Heidegger 1995, 76).

[5] MoE = Mann ohne Eigenschaften.

## The Shape of Modernity and the Question of the Postmodern

It was Fredric Jameson who recently characterized the postmodern moment as a crisis of representation. While intimations about such a crisis appeared in criticism in the late 1960s and early 1970s,[6] what is new, when we read Jameson, is the sense that it is thoroughly embedded in the presuppositions of west European consciousness. Following Heidegger's notion that "vor-stellen" (to "put before") and "per-cipere" can be equated,[7] Jameson outlines how imagining a "putative object" organizes that object for perception, with the proviso that "the object did not previously exist in that form at all" (Jameson 2002, 46). This is not just the idealist's claim that an object is reducible to our "idea" of the object. In the absence of a perceiving subject—a subject this argument in the tradition of Descartes is designed to bring about—it means that a subject borrows back its existence from objects imagined in this way. It is the manner in which we imagine objects, in other words, that allows us to construe the nature of the subject, and not the other way round. Our interest must therefore be drawn to the way in which we have arranged our imagination—or the representations our imagination brings forth—rather than to the question of subject and object per se. Since the type of imagining that Heidegger and Jameson describe for us has its origin in the construction of single-point perspective in painting since the Renaissance, it becomes apparent that perspectival seeing has brought objects before us in particular ways, and further made us aware that we are only ourselves as we are because we have learned to see in these particular ways.[8] As Jameson points out, this "does not mean that the object in perspective is merely a figment, an idea for me, a projection or product of my own subjectivity. It merely offers a certain construction of the real among other conceivable ones . . ." (47). The perspectival imagination arranges objects before us, as Descartes then puts it, in "clear and distinct" ways, although we would never recognize them as such if they were not arranged for sight in this manner. Since the purpose of the arrangement, by this argument, is to construct certainty (if not necessity) and to dispel doubt, our investment in the type of seeing we have arranged greatly increases. As a by-product,

[6] Typical in tone for such writings is Lefebvre 1991.

[7] The argument is unfolded in two speculative pieces Heidegger published in 1950 in a collection of essays under the title *Holzwege*. For an English translation, see "The Age of the World Picture" and "Nietzsche's Word: 'God is Dead'": Heidegger 2002.

[8] The most significant analysis of the role of perspective in the Renaissance was undertaken by Erwin Panofsky (1991), whom Heidegger undoubtedly read, originally published as "Die Perspektive als 'symbolische Form'" in *Vorträge der Bibliothek*. Warburg 1924–25.

this arrangement of our seeing will give us our modern idea of science. It also gives us the capacity to conceive of ourselves as subjects who see out into the world systematically. Our systematic perspectival seeing, in turn, inaugurates a notion of ourselves as sovereigns of knowledge whose acts of the imagination are directed at achieving certainty. Jameson then concludes: "It is only by way of this newly achieved certainty that a new conception of truth as correctness can emerge historically; or in other words, that something like 'modernity' can make its appearance" (47).

We enter "something like 'modernity'" with an imagination that will enliven objects for us by a particular process of representation and with an implicit account of ourselves to which the objects, seen in this way, will accord. Neither necessity nor substantiality accrues to the objects we apprehend in this manner. Both the objects seen and the manner of seeing are contingent, which is to say, the objects of our imagination could always be otherwise.[9] Perhaps it is a conviction about our own "lateness" in the era of modernity that encourages us to conceive of this moment of "double contingency" (that is, observations that "provide grounds for including contingency in meaning" [Luhmann 1998, 47]) as a recent event, yet surely Hume, with a healthy dose of Scottish skepticism, sensed such a moment when he digested Berkeley's new theory of vision, and Kant sensed it again when he read Hume's abandonment of rational causality and thereby awoke, as he put it himself, from a "dogmatic slumber."[10] I follow Jameson entirely on this point: modernity has been fashioned from such thoughts of contingency; we would not be in the age of modernity without them. At the same time, we must trace the arrival of the modern back to the moment when the perspectival imagination was inaugurated, and the era of modernity to the moment when we reached awareness that the arrangement of our seeing would show us a particular world, and not others. The first moment, prepared by the ancient Greeks, occurred during the early fifteenth century and gave rise to that flowering of humanism known as the Renaissance.[11] The second moment, prepared by Berkeley and Hume, can be found in Kant's insight that the

[9] This is also Niklas Luhmann's definition of contingency in his *Observations on Modernity*. (1998, 45), originally published as *Beobachtungen der Moderne* (1992).

[10] As Kant freely acknowledged in the preface to the *Prolegomena* (Kant 1990, 3:118). In a letter to Garve of 21 September 1798, Kant described the problem he set out to solve in the *Critique of Pure Reason* as a conflict of reason with itself: there is freedom in human beings, on one account of reason; there is no freedom, only the necessity of nature, on another. See T. E. Wilkerson's discussion of this point (1976, 116). Kant introduces the problem of freedom in the "Elementarlehre" of the *Critique of Pure Reason* (1990, 3:488–92: A 532–37/B 560–65).

[11] For Lefebvre, retracing attempts to conceive of a "science of space," the Renaissance also constitutes a turning point: Lefebvre 1991, 25.

transcendental imagination alights not upon "things in themselves," but upon "appearances" that this imagination interprets according to the manner of its own operation. With this interest in the nature and disposition of the subjective imagination, Kant gave not only the philosophical but also the social movement of Enlightenment in Germany its theoretical basis. In Kant's philosophy, then, the transcendental imagination is released from the objects it apprehends as a result of its representations. In doing so, the imagination can be observed as the initiator of representations, just as the representations will enable us to make deductions about the constitution of the "putative" subject that makes them. The shift away from intuitively grasped objects to an interest in "appearances" that tell us about our intervening intuitive understanding (but no longer indubitably about things in themselves) is the moment, we might say, that begins the era of modernity.[12] Postmodern theorists have recorded this moment, variously, as the end of the "classical episteme" (Foucault) and the end of "old European society" (Luhmann).

Of course, the psychological investment in certainty that Descartes began by no means ceases at this point. Kant therefore deduces a "unity of apperception" in the operation of the categories of subjective imagination, which is to say, a unity underwritten by rational precepts. Since he has glimpsed the chasm separating the transcendental imagination from the unknowable objects,[13] Kant must close this chasm with a conviction about a rationality that he holds to be a common feature of humanity.[14] So Kant dodges certain conclusions suggested to him by his method. He does not conclude that we are adrift on an ocean of our own imagining,

[12] The gulf separating modern times from intellectual positions assumed in antiquity, where the outer form of nature, but not its material substance, was contingent, is underscored by Kant (1990, 2:546: A 617/B 645).

[13] Kant considers whether the nature of the "systematic unity" of reason is subjective-logical, or objective and necessary, before concluding that it is an operative principle of reason that must be assumed—and no more than this. Nevertheless, Kant appears reluctant to sacrifice this principle's claim to objectivity altogether, arguing: "In fact one cannot see how a logical principle of the rational unity of rules could occur were it not for a transcendental principle through which such a systematic unity attaching to objects themselves were assumed a priori to be necessary" (1990,2:569–70: A 650–51/B 678–79). It is in such words that we detect Kant's disinclination to pursue conclusions forced on him by the logic of his own "Copernican" turn toward the subject.

[14] In considering the antinomies of pure reason, Kant seeks to show that "a path to certainty" still lies open to reason (1990, 2:409–12: A 420–25/B 448–43). These claims about reason must be weighed against other passages in the *Critique* where Kant acknowledges limitations to the rational enterprise. Here is but one example: ". . . reason has . . . only one systematic unity in the sense that it seeks to approach an *empirical, possible* unity, without ever reaching it completely" (1990, 2:512: A 568/B 596: emphasis in original).

as Nietzsche will conclude on the same premise.[15] Rather, he makes the assumption that our representations are "regulated" by the operation of our reason[16] in *practice*, thus shifting the ground underlying theoretical reason to a new plane of so-called "practical reason,"[17] where, among other things, it is our moral concepts that will be charged with evincing both the idea of the deity and the ontological investments we derive from it. The future operation of our representations will show us that our representations have been generated according to innate reason, and, further, that we can have certain knowledge that we are rational subjects because our representations have been generated in this way. However, this argument, which has only been unmasked fully in recent times, is a tautology—Nietzsche was the first to liken it to someone who has hidden a thing somewhere and knows in advance where to look for it.[18] The argument turns the regulative principle involved in the generation of our appearances into a constitutive principle used to explain them back to ourselves.[19] Nothing, however, allows us to make a justificatory statement out of this use of reason. By the same token, nothing prevents us from making representations that we hold to be rational. This indecision attaching to the rational content of the appearances we make visible to ourselves through the activity of representation—the question of whether our age has gone as far as it can with the rational idea or, as suggested by Habermas, whether it is still to be considered an "unfinished project" (Habermas 1981)—has become a prominent issue of the modern era. It is one of the main questions that exercised the mind of Robert Musil.

---

[15] Kant's failure to derive a compelling argument for the existence of God from either empirical or transcendental presuppositions leads to the sense of our being metaphysically unbounded.

[16] For Kant's conclusions about the regulative use of reason, in view of reason's failure to constitute anything beyond itself, see the third main part of the "Transzendentale Dialektik": (1990, 2:544–48: A 614–20/B 642–48).

[17] The shift from theoretical to practical reason is discussed—this is significant—only toward the end of the first critique, where it is adduced as a result of the need to move beyond assumptions about the unprovable existence of the deity to a position that assumes the deity as a "practical" necessity (1990, 2:556–63: A 631–42/B 659–70).

[18] Nietzsche sets out his objections to Kant's philosophy in a number of writings, but no more clearly than in *Beyond Good and Evil*, where he outlines the problem of the will in conditioning what we consider to be objective causality (and thereby the real nature of what we apprehend): cf. Nietzsche 1968, vol. 2, part 6, 18–20, 50–51.

[19] See Wellmer 1990, 46. One must, however, concede that Kant was aware of the danger of conflating a regulative principle with a (self-)constitutive one, but he held that this conflation, this "transcendental subreption," was actually unavoidable: cf. Kant 1990, 2:547–48: A 619–20/B 647–48.

## Critical Approaches to Musil's Modernity

In pursuing this question, Musil, a trained engineer, sometime inventor, and psychologist who had written his doctoral dissertation on Ernst Mach (see discussion of this work in chapter 2 of this volume), made a strategic decision to abandon the positive sciences. It was the crisis in the natural sciences at the turn to the twentieth century—a crisis precipitated by Nietzsche's attacks on rationalism, the discovery of general relativity, and Planck's uncovering of "quantum packets" at the atomic level of matter—that pushed Musil into literature, and the chance event of unanticipated literary success with his first novel (which he wrote after hours during an internship in a laboratory in Stuttgart) that kept him writing. In this first novel, *The Confusions of Young Törless* (1906), Musil has young Törless read, uncomprehendingly, Kant's *Critique of Pure Reason*. Musil thereby thematizes his own crisis about faith in reason, yet—and this becomes characteristic of Musil's approach to the question of rationality in general—he does not abandon reason altogether. The return of the—patently still rational—aesthetic subject enlivens the fiction of Musil when he might otherwise have abandoned the pretense of rationality altogether.

It is this turn toward the rational subject at a moment when the literary avantgarde in Vienna, to which Musil was loosely attached, had exercised the most profound suspicion about the Enlightenment project of reason,[20] that makes Musil compelling as a poet and important in the context of questions of literary value. I believe Musil, like his coeval Heidegger, fully comprehended the aspect of double contingency that, as Jameson argues, is latent in our notion of modernity, that is, our awareness that our representations, disjoined from things in the world, are not stabilized by any theory of how we make them. This was already the source of Törless's confusions in the first novel, and it becomes one of the main insights of the man consciously eschewing qualities in the later novel. It is also the guiding assumption of the Austrian dual monarchy. Of all the characters who come together at the behest of Graf Leinsdorf to brainstorm the "parallel action," Ulrich understands like no other the groundlessness for doing so. The interest of this society is to organize its representations not with regard to an outer reality it serves, or wishes to comprehend, but only with fidelity toward what it takes to be its own constructs. Hence even future events, such as a foreign regent's jubilee, are in principle copyable and repeatable, if these can already be found

[20] For Calinescu, the reaction of the avantgarde to the modernity of the Enlightenment from the middle of the nineteenth century, beginning in France, was concerted enough to constitute a rival modernity based on "extreme aestheticism." From this perspective, Musil would appear to take up a middle position between the bourgeois Enlightenment of the eighteenth century and the antibourgeois modernity of the new literary avant-garde (Calinescu 1987, 44–46).

within its own "event horizon." This is the sense of "semblances occur," the title of the second part of the first book of the novel, a title that both marks the autonomy of aesthetic appearances and celebrates it. In the absence of any Humean "common sense" about how we might live with the impressions we have become accustomed to (where common sense, say, would be held to be a common attribute of humanity[21]), the doctrine of "semblances occur" now lays down that our representations, however composed, become our reality. As Musil's irony about turn-of-the-century Austrian society makes clear, a notion of unattainable reality has been exchanged for an idea of attainable unreality. One might say, therefore, that Kakania upholds the Kantian moment in conscious fulfillment of the terms of contingency in modernity. This is the moment of separation of the transcendental imagination from things in themselves, a moment completely given over to the rule of aesthetic appearances with nothing outside the self to temper them. In a cultural moment Musil holds to be irreversible, the insight that there is nothing beyond the ideas of the imagination has become unchallengeable.

Stefan Jonsson and Albert Kümmel put forward two responses to this cultural moment. For Jonsson, the question that animates his study is whether the end of the anthropocentric point of view implied in this moment—the idea that the world truly "is" as we see it—also ends the humanistic hope for a better society. If subjectivity is diminished or decentered, do the prospects for social renewal, which had hitherto rested on certain assumptions about sovereign human agency in the world, diminish with it? For Kümmel, on the other hand, the conditions of modernity have inflated the choices available to subjects in spectacular ways. As subjects face the rising social complexity that accompanies the exercise of choice under the terms of modernity, the problem he identifies as Musil's main concern is that the understanding of subjects has not risen in comparable ways to meet and manage such complexity. Thus, where Jonsson speaks to the virtues of multiple identities, plural selves, and heightened levels of social choice, Kümmel already acknowledges the possibilities open to the modern subject, but looks to the technical means modernity makes available in constraining and guiding the "input selections" of subjects and citizens. In both cases, we move far beyond the traditional view that Musil's question of "the right way to live" is answerable in terms of the expressivist paradigm, that is, in terms of the view that the

[21] Despite the skeptical strain of his thought, Hume's philosophy was still governed by the humanistic outlook typical of eighteenth-century Europe. As Luhmann argues, the era of transcendental philosophy, to which Hume's thinking gave rise, especially in Germany, "may have been Europe's final attempt to achieve, with the withdrawal to one's own subjectivity and its consciousness, an order of knowledge that obligates cognitively, ethically, and aesthetically" (Luhmann 1998, 38).

representations of the artist as artist can meaningfully inform our actions in the world. Since they cannot, and since subjectivity, as conventionally conceived, is not likely to provide adequate responses to the postmodern moment that has become manifest within the situation of modernity, Musil must think his way into a new paradigm in order to address the problem of how to live.

For Jonsson and Kümmel, then, the rejection of the artist's frame of reference has to be assumed. Since this frame of reference arose from a notion of expressivity traceable to Romanticism at the start of the nineteenth century, the arsenal of expressive means open to the artist, Jonsson suggests, will not have a central place in making evident the sort of subjectivity we should follow in the future. Kümmel, attaching importance to the fact that Musil's protagonist is a mathematician, is even clearer on this point: the modernity that increasingly inserts itself into Musil's fiction as an operative idea is that of "a mathematical modernity which for its part divides into a counter-modern current whose origin is derived from an emphasis on intuition and a modern current whose mathematical language aims at a separation from all points of reference [to a real or given world]."[22] Musil will increasingly look to this non-referential mathematical modernity in constructing his alternative subjectivity. The innovation of Musil's novel, therefore, is an intimation of limitless possibilities on the level of subjective consciousness, where the narrative eschews the organizing principle of the imagination that represents things pictorially. Roger Willemsen (1985) characterized this innovative feature of Musil's fiction in the early 1980s as a narrative gesture that moves "gegen den Augenschein," that is, against the evidence of what can be visually represented. Kümmel makes the same point in a different way: "Not a novel *like* a wiring board (*Schaltplan*), a novel *like* a program. What is intended is the opposite. A novel *as* a wiring board" (2001, 111).[23] Both Jonsson and Kümmel, then, find the "utopian source" of the novel and of the "Other Condition" described in book 2 of the novel in the non-evidential categories of a new subjectivity, in, as Jonsson puts it, "the as yet unrealized possibilities of the present moment of modernity" (95).

In Jonsson's reading, we stay within the logic of modernity, for we have not exhausted its possibilities. "Ulrich," he tells us, "becomes an example of a human subject fit to meet the demands of a truly modernized world" (Jonsson 2000, 95). Does this deliver a sense of modernity as

[22] Kümmel's original German (2001, 68–69): "eine mathematische Moderne, die sich ihrerseits wieder in eine gegenmoderne, die Herkunft aus der Anschaulichkeit betonende, und eine moderne Strömung, die die mathematische Sprache von allen Referenten trennen will, aufspaltet."

[23] Kümmel's original German reads as follows: "[Nicht] ein Roman *wie* ein Schaltplan, ein Roman *wie* ein Programm. Gemeint ist das Gegenteil. Ein Roman *als* Schaltplan/Programm."

an "unfinished project"? In Habermas's conception, it is the creation of a self-consciously rational public sphere in the last quarter of the eighteenth century that gave modernity its decisive initial momentum. Modernity, for Habermas, arises above all out of this rational trajectory. For Musil, as Jonsson argues, it is different. On the one hand, Musil's protagonist is a rationalist drawn to scientific thinking: "[Ulrich's] skepticism dismisses the inflated ideals that contradict the findings of reason" (259). We need reason, in other words, to combat unbridled enthusiasm. On the other hand, the utopian quality of the novel can be found in an idea of shapelessness that subjects are called upon to work and manipulate in individual ways. This is the "utopia of human shapelessness" (*Utopie der menschlichen Gestaltlosigkeit*) on which book 2 is premised. This "shapelessness" does not issue from reason or any variant of it. Rather it has to do with the form of modernity that Musil now defends, precisely *against* the claims of privileged reason that had enlivened the eighteenth century,[24] and now also against the determinist ideologies of "gender, culture, race, nation, and empire" of Musil's own day.

Jonsson does not hesitate to call the "disposition of negativity" and indeterminacy that carries the novel an ethic of foundational significance. It turns every fixed form back on itself; it makes the emptiness of Kakania, which Musil had initially subjected to irony, into a crucible of new possibilities. Indeterminacy thus becomes a virtue. It dissolves the historical problem that had exercised Musil and for which there had seemed no answer—the problem of the "impossible choice between madness and fascism" (247). "A man *without* qualities," Jonsson explains, "equals pure subjectivity, that is, an anarchic madman or a criminal. A man *with* qualities equals pure identity, an individual drilled to become an obedient soldier. . . . Musil . . . posits the two poles as logical contradictions of each other" (246–47). Jonsson thus captures the postmodern moment of indeterminacy in modernity and valorizes it. It is the moment when the subject precisely resists the terms of the older subjectivity that had enabled it. Musil's novel is located, then, at a point of historical bifurcation of old and new. Since the expressivism of the older paradigm will not serve subjectivity, assumptions about the artist that had informed it must equally be rejected. Instead of the artist's sensibility, the subject becomes a political citizen, arriving at input selections in accord with the need to turn back the fateful determinations of history: the demagoguery of empire, fascism, and the habit of obedience of the soldier. Jonsson's global subject of the future transcends these fateful determinations in

---

[24] As Musil says in the novel: "The art of elevation above knowledge must be practiced anew" (*GW* 2:826). In the original: "Die Kunst der Erhebung über das Wissen muß neu geübt werden."

fashioning a postnational, as well as postmodern, ideal of self-consciously unstable and programmatically unfixed identity.

This political focus is completely missing from Kümmel's account. Instead of the intractability of fascism and madness, it is the emergence of modern knowledge systems to which Kümmel accords importance. Like Jonsson, but for different reasons, he assumes the decline of "optical certainty" under the Cartesian paradigm. The historical moment he disinters, however, privileges the technical discussion about knowledge, not its epistemological foundations. When old European society began crumbling away after the Thirty Years' War in the seventeenth century, it was above all the experience and training of war that got hard-wired into the European psyche: "Only he who controls his emotional feelings was in a position to obey—the laws of reason no less than the orders of an officer" (Kümmel 2000, 37). Since the question of knowledge shifts from an interest in obtaining certain knowledge to the manner in which knowledge is applied in practice, Kümmel can recast the contingency problem of modernity, much as Friedrich Kittler had done before him (Kittler 1990). In Kümmel's reading, Musil follows a double strategy. On the one hand, he adduces the notion of a "thinking of possibility" ("Möglichkeitsdenken") in order to proliferate the moments of contingency from which modern life is composed. This is the aspect of Musil's fiction that Musil criticism has always found compelling: the celebration of the free urban spaces of the modern metropolis. On the other hand, Musil restricts such contingency by a new technical idea derived from "mathematical modernity": the idea of the disciplining and regulation of the spiritual economy of the self, which will be explored in the final chapter of this volume. The provocative new idea is that this "MoE-Program," as Kümmel calls it (referring to the title of Musil's novel), this new notion of technical exactitude, is common to the protagonist and to the society of Kakania. Since they are patently not in perpetual alignment, however, Kümmel accounts for the disjuncture between the increasingly utopian strategies attaching to the main character and the increasingly dystopian commitments of his society by referring to Musil's use of irony. In the absence of an authorial standpoint that would render the jarring effect between the two levels interpretable for the reader—for the reader must be conceived as "disciplined" in the same way as the novel and its characters—Kümmel identifies a subversive strategy of "regulated non-functioning" (13) as the level on which the poetic intention of the text is realized.

How, then, does value accrue after knowledge is revealed as an arrangement of proliferating contingency (an arrangement that could always be otherwise), and criticism as the office that invests in the certainty of a now displaced, older knowledge paradigm, itself founded on a view about the sovereign status of criticism? Do we slip inexorably from groundlessness to pointlessness?[25] Do we lose interest in the literature that gave us our point of orientation? I have endeavored to indicate that the end of the older knowledge paradigm that made us invest in certainty as an epistemological condition of modernity was built upon a certain notion of the representation of subjects. Kant, perhaps, gave us the best definition of how representation would organize and construct appearances in view of the end of "things in themselves," even as he also gave us a now outmoded conception of the critical rationality that would inform our arrangement of the appearances and return certainty to us. Yet, as Musil shows, if we give up our idea that the appearances can be made transparent to us by the rule of the aesthetic imagination, grounded in notions of the autonomy of the subject (or in an underlying morality that informs it), we do not automatically lose every idea that we associate with modern life. For we remain political subjects who continue to make real political choices—choices that matter in our immanent lives—and we remain individuals who have recourse to the technical means we have refined in the development of our knowledge systems. More than anything, as Jonsson and Kümmel show, we move beyond an idea of the sovereign subject to a new idea of the individual as singularity. As a singularity, in this "singular modernity" (Jameson), it is contingent experience that shapes the individual, not any version of transcendent subjectivity that allegedly precedes it. For as even Kant tells us, there is nothing we can find in the idea of the imagination and the reason supplying its concepts that will give us final certainty about the assumptions we follow. We must cultivate reason as something provisional ("mit gutem Glücke gebraucht": Kant 1998, 3:579). The further challenge that singularity poses, then, is that our ethical commitments must be arrived at on a different basis from that which was used to give us our ontological ones. If we allow a disjuncture between our ontological and our ethical commitments, as Jonsson and Kümmel suggest, then we must look beyond assumptions about the centrality of art (or reflection on art) in life to give us orientation, just as we will fail to find any given "cultural capital" in our aesthetic judgments. Not even Guillory's provocation about an "aestheticism unbound" (340) will make access to the "game of culture" ultimately any less contrived and partial, any more moral and real. This already suggests the unyielding sense of our singularity and the experimental nature of our "postmodern" challenge, which urges us to tolerate, even celebrate, nothing less than the inescapable nature of our finitude in time and space. As Musil remarks in a diary entry (also noted by Kümmel):

[25] For Heidegger, the emptiness of modern culture and its "repugnant" psychologizing has turned boredom into a basic "attunement" of life. He identifies facing the problem of boredom as one of the urgent tasks of philosophy (Heidegger 1995, 166–67).

In a human life neither the facts nor inner conditions are variable but their ordering in space and time [is]. Individuality is a process, a variation. Finished with its death.[26]

[26] "Verschieden im Menschenleben sind weder die Fakten noch die inneren Zustände, sondern ihre raumzeitliche Anordnung. Individuum ist ein Ablauf, eine Variation. Fertig mit seinem Tod" (*TB* I:452).

# 6: Musil's Concept of Value

CRITICAL OPINION ABOUT *The Man Without Qualities*—and by implication, the works leading up to the novel seen to represent earlier stages of a consistent intellectual project—has not reached unanimity in regard to Musil's intentions. Disagreement is concentrated in particular on the meaning of the sibling idyll in book 2 of the novel. Part of the disagreement among critics has come from the fact that the novel was not completed during the author's lifetime. In this respect the reader of Musil's novel faces a situation of uniquely challenging proportions, where questions of this type must be raised: did the novel remain incomplete simply because Musil ran out of time at the end of his life, or was he unable to complete it because of his own uncertainty about its themes and direction, thus suggesting that Musil would never have finished the novel even if he had lived longer?

Since these questions have never been properly answered in Musil scholarship, some critics have preferred to focus attention on those parts of the novel that were published during the author's lifetime, in particular, Musil's satirical portrait of Viennese society that forms the focus of book 1 of the novel. The highwater mark of such attention is certainly Norbert Wolf's monumental *Kakanien als Gesellschaftskonstruktion* (2011), a study that values the novel for the comprehensive sociocritical portrait of modern life that it draws. David Luft's *Robert Musil and the Crisis of European Culture 1880–1942* (1980) had already assayed similar territory in an early study written in English. Another English-language critical work of note is Mark Freed's *Robert Musil and the NonModern* (2011), which arises from the interest in the sociological, but connects this interest to a critique of modernity (see chapter 5 for an extended discussion of Musil in relation to the crisis of modernity). Freed's study is valuable for several reasons, including for the attention it gives to the idyll of the sibling lovers. Freed reads this idyll, and the ground of the "Other Condition" in which it is rooted, in positive terms as "a strategy of autonomy" directed against those who would seek "to realize their wills through the instrumental use of others" (Freed 130). Freed's understanding of this strategy of autonomy stands on common ground with other notable approaches to the novel, such as that of Jonsson (2000),[1] insofar as Jonsson also highlights oppressive tendencies in the portrait of Kakania that

[1] Jonsson's study was discussed at length in chapter 5.

make urgent the search for alternative ways of imagining personal identity. If in early-twentieth-century Viennese society identity was governed by a process of "[gluing] a selection of approved identities of gender, class, and ethnicity to the subjects of Kakania" (Jonsson 13), then only a radical break with such a society leading to the conception of an entirely new human being—the idyll that is put forward for consideration in the second book of the novel—appears to offer a meaningful response to the social paralysis depicted in book 1. It seems, then, willy-nilly, that the reader is led back to the question of how the second book of the novel sits alongside the first, notwithstanding the fact that this question might not admit of a definitive answer on the basis of the published version of the novel, and not even adequately responded to when a sizeable amount of hitherto unpublished materials relating to the novel is included, as as is the case in the German edition of the novel since 1978. The availability of an even more comprehensive, digitized collection of these materials since 1992 has not changed this basic situation.[2]

The reader of Musil who aims at a full assessment of Musil's novel and his intentions must therefore accept an invidious starting position: we will never finally know how Musil intended to complete the novel; the published version of the novel is destined to remain a "torso." Indeed, it is not even possible to confine consideration of the finished parts of this torso to a reliable core, since it can fairly be assumed—Jonsson's and Freed's approach are examples—that any view of book 1 of the novel is to some extent dependent on what can be gleaned of the probable shape and intention of book 2. Here the position is by no means unjustified that Musil's longer-standing forward commitment to his materials in fact lay with book 2—the most open and least finished part of Musil's project. By the mid-1920s these materials were important enough to provide the title for the entire project: "The Twin Sister" (*Die Zwillingsschwester*). Musil sent materials pertaining to a novel with this title to Rowohlt, his publisher, at the end of 1925 and in early 1926, five years before the novel's eventual appearance in print at the end of 1930.[3] As Helmut Arntzen reports, it was not until the beginning of 1927 that Musil was to settle on a final title for the novel (Arntzen 41). It is also worth noting in this context that the drafting of chapter 116—the chapter that extends, and to some extent concludes, the discussion of what I am calling "the question of science" in this volume—was not commenced until a very late stage of the writing process (see Arntzen 43). If this is so, the conclusion appears warranted that it was Musil's thoughts on the problem of the two

[2] Musil 1992.

[3] See Helmut Arntzen's discussion of the gestation of the novel (1982, 30–50). On April 30, 1926, in an interview with the journalist Fontana Musil referred to the novel as having the title "Die Zwillingsschwester" (*GW* 7:939).

cultures—a problem he had been puzzling over at least since his doctoral dissertation on Ernst Mach—that was to give book 1 of the novel its final shape and that, in consequence, justifies extended consideration of Musil's discussion of science and his treatment of scientific discourses in his oeuvre. I focus attention on chapter 116 of Musil's novel in my conclusion to this volume.

If we now return to the problem of how book 1 of *The Man Without Qualities* sits alongside book 2, we must conclude that the reader of the novel is faced with a choice. This is a choice, urged upon the reader in view of the unfinished nature of Musil's project, that requires readers to establish in their own mind a level of coherence between book 1 and book 2. The reader must decide whether book 2 is meant as an answer to problems identified in book 1, and, if so, what type of answer. If we follow the lead given by sociocritical approaches to book 1 of Musil's novel, we see that these approaches tend to come together in the view that Musil's portrait of Kakania is meant pejoratively. In other words, most of the critical opinion about Musil's analysis of turn-of-the-century Viennese society, which many see as a prototype for modern society in general, takes Musil to be condemning, and not merely critiquing, this society.

The gist of the sociocritical novel—a story of a particular society, "Kakania," at a particular moment in its history, the period shortly before the outbreak of the Great War—is as follows. It must be noted upfront that "Kakania" is a disrespective label (suggestive of the baby word for "poo") derived from the administrative term "imperial and royal" (*kaiserlich und königlich*). The label indicates that it is an administrative elite, rather than the imperial rulers themselves, that keeps the sprawling, multiethnic, long-lived Austro-Hungarian Empire ticking over. It is this elite and its ruminations and interactions, accordingly, which absorb attention in book 1 of the novel. As the novel quickly suggests, however, Kakania has reached the pinnacle of its cultural development and has already—and perhaps long since—entered a state of decline. We know from the course of history (to this extent the novel makes a particular use of historical facts, even if Musil was right to declare it in other respects not to be a historical novel[4]) that this decline was to be understood as terminal. In the months before the outbreak of war when the novel begins, Kakania seeks to renew itself. The mechanism chosen for this task is a so-called "parallel action," the twinning of two cultural celebrations due to be held in 1918 marking the 70th year of rule of the Habsburg monarch and the 35th year of rule of the German emperor. As we know from history, this is the year that the "axis" powers Germany and Austria-Hungary were both

[4] In the interview with Fontana he declares the novel not to be a historical novel (*GW* 7:939).

swept away, thus ending the imperial rule of both the Habsburg and the Hohenzollern dynasties in Europe.

A committee is convened to plan this celebration at the house of the Ermelinda Tuzzi, a diplomat's wife and eighteenth-century-style society lady. The thirty-two-year-old protagonist, Ulrich, the eponymous "man without qualities," is welcomed onto this committee. Ulrich is by no means at the peak of his own powers and, indeed, has reasons to feel a private sense of failure, having found no satisfaction to date in various career pursuits, while at the same time remaining aloof from any cause or social concern and certainly any meaningful connection to his own family. In fact, Ulrich also soon finds himself in considerable disagreement with the parallel action and many of its committee members, who are drawn mostly from the professional and privileged classes in Viennese society. Kakania is an apt descriptor of this society, since it seems soulless in terms of its values, and directionless, when compared, say, with its more strident and outgoing neighbor Germany, which is represented on the committee through the figure of the industrialist Arnheim. Ulrich, striking an attitude of openness to all things and all possible courses of action, not excluding thoughts of suicide, seems himself scarcely able to resist the proliferating cultural malaise depicted in this part of the novel. How he might overcome his lassitude and gain a sense of purpose in life is one of the key questions brought to attention in the book 1 of the novel. Since, as mentioned, most critics agree that the portrait of Kakania and the parade of functionaries and personalities brought into view by way of the parallel action was ultimately meant negatively, both the war and the prospect of social dissolution that historical events foretell suggest a bad outcome for this society. Musil's annotations on the progress of the novel reflect this. In materials included in the longer version of the novel, Musil makes this observation: "Basic idea: war. All lines lead to war" (*GW* 5:1851).

If this seems broadly uncontroversial as a summation of book 1, it remains to be assessed whether and to what extent book 2 is a response to the circumstances of impending social dissolution book 1 describes. I argue below that the sibling idyll in book 2 of the novel, in general terms, was indeed meant as a response to the social decay of Kakania. But to stabilize the transition between books 1 and 2 in this way is only to begin the process of analyzing what might be at stake in the Other Condition and what Musil might have intended with it. The main question I attempt to deal with in what follows is whether Musil meant to endorse the unusual idyll that he constructs between the siblings and whether this idyll can be viewed as a constructive alternative to the events depicted in book 1, which seem destined either to continue fruitlessly or to end in some form of wider social fragmentation or outright collapse.

I begin consideration of this question by pointing to a core aspect of what might be called Musil's cosmology. This aspect concerns the importance attached to individual experience. To highlight the value of such experience in Musil's thinking, I consider Musil's brilliant short story "The Flypaper," first published in 1913 and reprinted in Musil's last published work, *Posthumous Papers of a Living Author*, which appeared in 1936. This story, crafted consciously as a type of modern parable, provides a way of understanding the kind of displacement of interest that occurs when a description of group behavior is the initial focus but is progressively abandoned as the perspective of the individual takes over. The displacement of the bird's eye view of events—a version of "the view from nowhere" typical of the scientific approach[5]—through the cueing of the importance of the individual is a characteristic feature of the author's mode of storytelling and was already present in his first novel, *The Confusions of Young Törless.* Consideration of "The Flypaper" helps us see, as it were in miniature, that the premise of singularity is the philosophical starting point for Musil's intellectual commitments as a whole.

## The Primacy of Singularity

An unnamed observer records the moment when a fly lands on a sheet of sticky paper designed for the purpose of trapping and killing it: "The flypaper Tangle-Foot is approximately 36 centimeters long and 21 centimeters wide; it is coated with a yellow, poisonous glue and comes from Canada. When a fly lands on it—not with a particular desire to do so, more through convention since there are already so many others there—it gets stuck first only by the outermost segments of its little legs" (*GW* 7:476). Despite the expectation of impartiality aroused by the allusion to facts at the start, the observer's emotions quickly take hold, as the presence of something "harrowingly human" appears in the midst of a process of decline that unfolds before this observer. The insight is soon registered that humans also seem stuck fast, are also enchained to something they cannot escape, and that all life—human as well as animal—is inescapably tied to a single brutal truth: the fact that death[6] is final, non-negotiable, and perpetually present. "The Flypaper" starts, therefore, with existentialist assumptions about the certain annulment that all life forms confront. This insight already provides a link between the fate of the fly and the gaze of the story's narrator—a gaze whose claim to constitute a version of the general or the scientific is sustained throughout and increases at the end of the story in the form of the technical contrivance of the

[5] See Thomas Nagel's discussion in his immensely valuable text *The View from Nowhere* (1986).

[6] For Musil, death is the conditio sine qua non of creation and of perception, and fear of death is the conditio sine qua non of ego-emotion (*TB* 1:140).

magnifying glass (I considered the type of seeing associated with this gaze in more detail in chapter 3 of this volume).

The proto-scientific or pseudo-scientific gaze of the observer records each moment of significance in the capitulation of the fly to its uncomprehended opponent. Just as the fly's contortions on the flypaper only add to the coating of poisonous glue that sticks to it, hastening the process of decline, so human beings also resist, equally fruitlessly, the inevitable fact of physical decline. Just as the fly attempts to make light of its slow demise, so it is with humans—the tabetic sufferers who seek to maintain the appearance of normal comportment despite the fact of their wasting bodies, or the clapped out old military men, who walk with bowed legs as if they're negotiating a narrow ledge or a balance beam. As the intervals between resistance and the need for respite lengthen, it is the process of succumbing to a mortal foe that commands the narrator's interest, though death itself is not the subject of any intrinsic interest—the narrator calls it "a nothing, an it." Of greater significance is the life force that intensifies in the fly in the last moments of its struggle. It is in these moments that the singularity of the fly's perspective begins to appear. The fly that had landed by following a convention, by behaving as a member of a group, now dies a very individual, even lonely, at any rate a singular death. The narrator, surveying the last moments in the life of the fly through a magnifying glass, sees "a small flickering organ . . . like a tiny human eye that ceaselessly opens and closes" (*GW* 7:477). In this final description, the tiny eye of the fly has been conjoined to the eye of the observer looking through the magnifying glass. In these last moments, the fly and its observer become one.

This apparently unremarkable insect thus attains the status of true singularity by the end of the parable. The narration, which at the beginning seemed committed to a general view of species behavior through a stance of cool impartiality, now registers what can only be construed as the obverse of these initial intentions, namely a touching insight into the fly's "subjectivity," its otherwise unseen inner life, the parts of its struggle that resemble, and ultimately become equivalent to, the moments of struggle of human individuals with their own fated demise. The narrative perspective, in this way, gives up its pretention to impartiality and finally sides with the fly—the fly as human, and by a perspectival reversal: the human as fly. The narrative perspective, by now thoroughly unscientific, thereby attains a certain humanism. The reader is led to an acquaintance not with the fact of death, about which nothing at all can be known or said, nor with any putatively "objective" knowledge of the stages of a process of physical decline, but with a life whose private inner struggle, otherwise unseen and unknowable, becomes the source of its meaning and its value.

If we move from "The Flypaper" to the sociological questions we have already identified as present in *The Man Without Qualities*, we can see how the initial interest in Kakania is a prelude to the real interest—the view through to the individual. Since the general interest is aided and abetted by the scientific attitude—"the view from nowhere" that science works from and seeks to promote—we may also say that the problems associated with general understanding are inherently also those associated with the scientific interest. Any trumping of the particular by the general—the charge that will be laid at the door of the totalizing tendencies of Kakania—will therefore also be a charge laid, at least initially, at the door of science. Sorting out what is good and desirable about science and scientific procedure from what is undesirable and "bad" will be a question of interest in book 1 of the novel. That Musil might have come to this question fairly late in the gestation of the novel, despite his long-standing personal interest in scientific questions, means that the question of science might be considered a kind of overlay or non-essential feature in the novel. Our consideration of "The Flypaper," however, in fact reveals that the question of science is implicated in the form of the general itself.

Following the "outside-in" perspective established in "The Flypaper," it is evident in *The Man Without Qualities* that generality, predicated on the primacy of the universal over the particular, has swamped the particular feeling that individuals maintain toward their lives. The individual, Musil's protagonist Ulrich, takes up resistance against mechanisms that seek to impose patterned, prefabricated forms of identity on him. Justification for the strategy of resistance that he adopts lies with the encroaching sense of dissolution that this individual witnesses in his society. But of course resistance is only the beginning. If, as history tells us, Kakania will soon succumb to the entropy it cannot successfully forestall, how is the individual to turn away from his society and gain release from the forces of terminal decline that everywhere press down upon him? How is a genuine autonomy of the self to be found and secured against the encroachment of prefabricated identity that Kakanian modernity has turned into such a virtue?

The answer to these questions is to be found in book 2 of the novel, and specifically in the Other Condition—the fertile ground on which thought and feeling meld and from which the siblings construct the private utopias that attempt to break with the strictures of their society. The suppositions that inform the Other Condition are derived, as I will argue in the following, from the philosophy of Friedrich Nietzsche—for at least three reasons. Nietzsche's philosophy is of particular appeal to Musil, first, by virtue of its orientation toward the fact of singular experience; second, because it is constructed on existentialist assumptions that underscore the ever-present "levelling" fact of death in line with

Nietzsche's "death of God" thesis; and third, in a post-religious context, because its energy and dynamism as a philosophy are directed at developing strategies of release if not outright escape—escape from nihilistic life-denying forces on one side, release into authentic, life-affirming and -enhancing experience of being on the other. Since Nietzsche's significance for Musil has been subjected to some doubt in recent years, particularly in Anglophone scholarship, I devote attention in the following section to the need to arrive at a balanced assessment of Nietzsche's influence on Musil. I ask in this section whether Nietzsche's philosophy truly provides the underlying motivation for the sibling idyll raised into view in book 2 of the novel.

## Evaluating the Significance of Nietzsche

Among the vast number of allusions in Musil's diaries and manuscript papers to thinkers the author consulted over the course of his career as a writer, no name is encountered more often, and noted with greater assiduity or in greater depth, than that of Friedrich Nietzsche. One pertinent quote among a great many that could be taken from Musil's writings is sufficient to underscore Nietzsche's general significance for Musil: "My formulation (*Formel*) of creativity as an ultimate value stems from N(ietzsche)" (*GW* 5:1777). More than anything else, it is the approach to moral questions and indeed the entire terrain encompassed by the moral in Musil's understanding of it that betrays Nietzsche's deep influence from the very beginning. In the critical literature on Musil it is even claimed that all of Musil's most important thoughts can be traced back to Nietzsche in one way or another.[7] As widespread as this view of Nietzsche's importance for Musil undoubtedly is, there are also dissenting voices, particularly in recent scholarship. Commentators such as Patrizia McBride (in *The Void of Ethics*, 2006), for example, acknowledge Nietzsche's prevalence in the author's thinking but dispute orthodox scholarly opinion about Musil's commitment to Nietzsche, especially in the major novel. So while there is no denying Nietzsche's influence as a whole over Musil's thought, the question that arises for these critics is whether Musil endorsed this influence on all levels. Did Musil, in other words, notwithstanding a clear intellectual debt to Nietzsche, turn against Nietzsche in the final section of his novel when he was formulating key aspects of the sibling idyll? And is book 2 of the novel finally to be read as anti-Nietzschean?

The question of Musil's use of Nietzsche in the novel *The Man Without Qualities* is not just a trifling question about how one should read the use of sources in Musil. It also touches on a wider dispute about the

[7] See especially Dresler-Brummer (1987).

nature of the modernity we inhabit,[8] how this modernity is premised, particularly in terms of its ethics, and whether the stance of openness to possibility that Musil outlined in book 1 of the novel—what Mark Freed calls the "interest in counterfactual possibilities" (Freed 50)—is to be welcomed or lamented. Nietzsche figures at the center of all these questions. The dispute about Nietzsche in Musil highlights, I believe, a basic question about what we make of the ethical openness we ascribe to the paradigm of the modern and whether it really helps the human being *at a values level* to indulge in the variety and the sense of possibility that characterizes modern life. McBride's case against Nietzsche in this respect is instructive, for it is Nietzsche—by way of the particular reading of Musil that is advanced—who appears to be convicted of peddling false hopes that do not finally bring about any ethical advancement in the human being. If Musil is seen to fall in behind this view, then at the very least, Nietzsche's importance for Musil, as established and defended by numerous studies in Musil scholarship, would have to be fundamentally revised. Any revision of Nietzsche's standing for Musil would also have implications for the way we should understand Musil's ethical project as a whole. Since McBride is a serious and intelligent reader of Musil, we must take time at this point to examine the argument she puts forward.

McBride's argument, as I read it, has two parts. As the stronger commitment in her argument relates to modernity, I will review this aspect first. This is how McBride frames her general view of the modern as a cultural paradigm: "This study offers a reading of the Other Condition as Musil's dramatization of an ethical void built into human existence" (McBride 18). That such an "ethical void" exists is a cultural-philosophical conclusion that emerges from McBride's reading of Lyotard and particularly Agamben. A footnote reference reveals Agamben's influence on McBride's argument. Agamben is interpreted by McBride as viewing Musil "as the epitome of a critical project that investigates the void which opens up when thought confronts the impossible scission between poetry and philosophy, ecstatic-inspired language and rational-conscious thought" (McBride 172). The first part of McBride's argument, accordingly, endorses this culturally pessimistic view that a void is to be found in the form of human existence that has become established in the era of modernity.

The second part of McBride's argument is derived philologically, though I believe the philological conclusions McBride reaches are taken over from the culturally pessimist assumptions that she takes over from Agamben and others. McBride says two things at this point. The first

[8] Questions about the current shape of modernity were also raised in my previous chapter. How rational this modernity is—or should be—is a central consideration in that chapter.

thing is that Musil changed tack in his understanding of modernity "in the 1920s and the 1930s" (16). Building on assumptions about the apriori nature of the "ethical void" in the modern condition, McBride argues that it was Musil's intention "to dramatize the failure of Ulrich's attempts to ground his life in [the] utterly moral condition" he called "the Other Condition" (17). Since the Other Condition is seen in Musil scholarship as the ground from which a series of ethical utopias and an opposition to the emerging fascism in Musil's day are launched, McBride must turn her argument against the prevailing critical orthodoxy. She does so with this argument: the positive valorization by older critics of the Other Condition in the second (unfinished) book of the major novel occurred because these critics were not able to take account of "reliable and comprehensive editions of both the novel and the notebooks, which appeared in 1976 and 1978." As she argues: "Previously unpublished materials included in these newer editions show that the development in favor of the narcissistic 'utopia of the Other Condition' had been planned at least since the mid-1920s, a circumstance that makes a reading of the Other Condition as Musil's resigned response to fascism difficult to sustain" (17–18). With the philological supports for this older view of Musil thus weakened, McBride establishes her own reading of the Other Condition as a failed attempt to sustain an ethical vision and "a warning against the dangers of overdetermining the ethical by filling its cognitive void with yet another absolute vision of the good life" (19). In order to close this ethical void and put something more salutary in its place, McBride outlines the presence of a "Kantian horizon . . ., a reservoir of heuristic categories and productive images" (20) derived from Kant's third Critique, in particular "Kant's discussion of the disinterestedness of aesthetic pleasure" (109) in the *Critique of Judgment*, "that frame[s] the insights Musil derived in part from his scientific studies" (20). In McBride's view, then, Musil is not an acolyte of Nietzsche at all, or at any rate does not remain one, but turns into a Kantian in the last years of his life as he attempts to finalize the major novel.

The argument McBride advances rests, I believe, on two missteps. Clarifying them will lead to a better appreciation of Musil's true intentions in the major novel. It will also lead us to assess what it is at stake when arguments of this kind are advanced in scholarship.

The first misstep concerns the nature of the Other Condition, which McBride sees as promoting a series of false utopias. Against McBride I maintain the opposite viewpoint—that the Other Condition, as Musil intended it, is a fertile ground from which positive forms of ethical thought emerge and on which they are predicated. The subtitle of this second book of the major novel makes this point, though there is a trap here for the unwary. The subtitle of this section is "die Verbrecher," a noun that is usually translated into English as "the criminals." That Musil uses this word in a special sense, however, is suggested by the fact that it appears in parenthesis. That there might have been a need to set a trap for the guileless or superficial reader of the novel emerges from the political background at the time of writing—the novel's second book was published in Germany in 1933, the year of the "enabling act" that secured Hitler the executive powers needed for a political dictatorship. Thus, as I detail further in the section below, both the title and subtitle of book 2 are meant ironically. But one easily sees how such a title could predispose the reader to view the idyll of the siblings Ulrich and Agathe as expressive of a negative intention.

On top of this, it is decidedly not the case that the notebooks, published in 1976, or new manuscript materials included in the edition of *Der Mann ohne Eigenschaften* released by Rowohlt in 1978, have the philological significance for Musil scholarship that McBride claims for them. McBride suggests that a clear line to her view emerges from these materials. But this suggestion is contentious and not sustained by a close examination of these materials. Instead of an emerging picture of the failure of the sibling idyll, one rather sees in these materials the great many difficulties that dogged the continuation and completion of Musil's expansive project in the last years of his life. These difficulties were only increased by the deterioration of Musil's physical environment in the 1930s in the face of declining income from his writing, growing alarm at the spread of fascism and anti-Semitism, and, finally, the fact of Hitler's annexation of Austria itself, which pressed the need on Musil and his wife to leave Austria in 1938 and escape across the border to the neutral territory of Switzerland. While some of these difficulties, it is true, might be considered to relate to Musil's intentions with his materials, rather more of them seem to concern technical questions regarding the crafting of complex materials and the need to keep the sense of wider fragmentation in the society Musil depicted separate from the development of the counter-ideal of the siblings that, as mentioned, was a long-standing forward commitment of the author. Overall, there is insufficient evidence from the unpublished drafts and other materials relating to the novel that suggests the kind of change in the direction of the Other Condition that McBride believes to have occurred in Musil's plan for the development of the novel in the 1920s and 1930s.

If we review the matter of Musil's intellectual influences as made evident in his diaries, we equally have little reason to incline to the view that, over time, Musil moved more toward Kant and away from Nietzsche. As McBride readily concedes (2006, 188), Kant is referenced sparingly in the diaries, is hardly ever the subject of close attention, and when his name is mentioned or implied it is mostly in non-substantive terms. When substantive attention is registered, for example in notebook 10 from the period 1918–21 (with some comments added in 1929 and 1939), Musil refers to Kant, somewhat dismissively, as a thinker in the scholastic tradition. By contrast, Nietzsche is a consistent presence in the diaries from the earliest entries to the last; one is never left in any doubt that Musil considers Nietzsche to be a thinker of high importance for his age. In the latter part of notebook 34, which was written in the period from 1930 to 1938 (so the entry concerned is likely to have been made in the late 1930s), we encounter this summation of Nietzsche's significance for Musil: "Did I absorb in my youth even 1/3 of Nietzsche? And yet decisive influence" (*TB* 1:903).

Since, on the evidence available, we cannot endorse McBride's view of Kant's *specific* importance for Musil on particular points of philosophy (though his general importance for the thought of Musil's age is certainly conceded and discussed throughout this volume), we can now move to consider what I hold to be a second misstep evident in McBride's argument. This is the view advanced by Agamben and others that Musil held to a culturally pessimistic position about modernity, involving the key insight that modernity has an "in-built ethical black hole." My contention here is not to ask the big question[9] about whether modernity really must be thought of as having this problem or failing. Maybe it does, as many cultural commentators of a postmodern orientation argue (for further discussion of the question of modernity's crisis, see chapter 5 of this volume). My disagreement with McBride is whether Musil held such a view. Let us therefore review the argument outlined in chapter 4 in relation to Musil's ethics. We will then be in a position to come to a better understanding of Nietzsche's importance for Musil.

In chapter 4 it was established that, for Musil, ethical thinking is properly encountered in a domain of thought or a type of practice of thought that Musil characterizes as the "non-ratioïd." This is thought referenced against the singularity of human experience—of, as Musil puts it, the ascendancy of the exception over the rule (and the habit of rule-following). Musil calls the contrasting domain of rule-following the "ratioïd domain" and allocates the scientific interest in general to this domain. Rule-giving and rule-adherence undoubtedly play a role in modern life, and to the extent that the pursuit of modern science plays a prominent role in modern life, rule-giving will remain important. But Musil equally considers this role exaggerated. It is exaggerated because a major area of modern thought relating to the ethical comportment of the human being is not typically encountered within the ratioïd domain—a fact that will strike the proponent of Kantian moral theory as surprising. Kant's moral thought both in the *Foundation for a Metaphysics of Civility* and his second major critique, the *Critique of Practical Reason* are constructed around the importance of a moral law whose various manifestations

[9] I considered some of the questions about modernity in the previous chapter.

hold the principle of rule-giving to be pivotal. If, according to Musil, the domain of rule-adherence (the ratioïd domain) is not where we must look in order to do ethics, it equally follows that Musil—I maintain under the influence of Nietzsche's thinking, which stresses the contingent and "constructed" quality of all ethical thought—must equally reject Kant's ethics. Given the widespread cultural prestige attaching to Kant's moral thought throughout the nineteenth century, any position that rejects Kantian morality is likely to appear, at the very least, at odds with the mainstream. This was certainly the case with Nietzsche, who cast himself in all his writings as "unzeitgemäß," as out of step with his time. In view of these considerations, all of which can be pieced together from a study of Musil's 1918 essay "Sketch of a Writer's Knowledge," one does not need to look far to see how Musil makes sense of ethical matters.

We can summarize Musil's position about ethics as follows: Musil concludes that Nietzsche, absorbing utilitarian arguments being advanced in discussions about ethics in England in the early to mid-nineteenth century, particularly in the thought of J. S. Mill, is right: ethics is not universal, it does not rest on a single "categorical" moral principle, nor is it to be imagined as lying exclusively within the gift of rational logic, as Kant could be taken to maintain, but rather has its own history and historical situatedness that can be linked to those who have social standing or power and who fashion a values hierarchy, as it were, in their own image (this is more or less the argument Nietzsche advances in the *Genealogy of Morals*). Musil takes on such historico-cultural assumptions informing the field of ethics, but then, beginning with his 1918 essay, radicalizes them in the direction of the individual whose outlook on life must be accorded highest significance: in the domain of the non-ratioïd, ethics must be determined according to the constraints, and with respect to the ideals, of singular human experience. So the demand Musil makes of ethics is that it take cognizance of the individual and the settings and situations in which individuality and singular experience occur. To come to an uncharitable view of this individuality and these settings, as McBride does, would be to dismiss these utopias of the Other Condition as "narcissistic." But this would make light of what, in book 2, are immensely complex discussions between Ulrich and his sister Agathe, a sister whom he thinks of as his twin, though she is five years his junior. Agathe is a "twin" on many grounds, not the least of which is the fact that she shares Ulrich's view of the non-negotiability of the singular perspective and the freedom (such as it might be) that it confers. But this does not make her into his double, for Agathe is also someone quite unlike Ulrich, someone who says things he would not say and does things he would not do. Book 2 of the novel accordingly presents the reader with a series of disquisitions between equals about the nature of individual ideals, in regard to which the discussants may well differ, not with any simple or uncritical endorsement of the ideals each might be minded to advance. One must conclude that the ethical work of the individual carried out in the region of the non-ratioïd is by no means a series of excursions in a garden of pleasures (some of them illicit), as some readers of Musil might consider it to be. At any rate, we do not need to look far for the source of the so-called "ethical black hole" in modernity. If we can say such a hole exists, it opens up in thinking not despite Kant but because of him, and precisely because of doubts about the social and philosophical prestige accruing to Kant's moral teaching that the novel expresses. As we have seen, Kant's demand is that ethics be looked upon as a matter of the application of a moral super-principle. Musil questions whether this kind of ethical thought can be defended in a modernity in which the individual attempts to cultivate, and follow as truly as possible, a deep sense of individual life and the ideals flowing from it.

On one level, views such as McBride's are suspicious of a basic conflict between a "bad modernity" that can be attributed to (a particular reading of) Nietzsche and a "good modernity" constructed from the allegedly solid material of Kant's aesthetic theories (it is a weakness of McBride's position that Kant's moral theory is ignored in favor of his aesthetic thinking, to which moral positions more cogent than those actually advanced by Kant are ascribed). The task at hand, it would seem, is to find the right form for such a modernity to save it from putative excesses, where it is assumed that the bad modernity arising from narcissistic utopias will ultimately end in the trauma of Hitler's fascism. We must accordingly take care to diagnose Musil's understanding and treatment of Nietzsche properly, but equally not make the mistake of anachronism that arises from the imprint of today's cultural preferences—as defensible as these may seem in isolation. This mistake would consist in separating Nietzsche from Musil in order to rescue Musil's outlook, against the latter's stated views about Nietzsche's influence, and, as a corollary to this position, attaching Musil to Kant because of Kant's allegedly "better" dualism of intellect and feeling (equally against Musil's merely modest interest in Kantian thinking). It may well be that Kantian dualism, and the apodictic quality of his moral thought, is a better bet today, culturally speaking, than Nietzsche's view of radical individuality and his redrawing of the boundaries of ethics. However that may be, the view we must come to, in line with the available evidence, is that Musil's utopian scenarios are premised on, and require endorsement of, the non-negotiability of the individual in all attempts to configure the ethical. That the collective commanded an ever larger role in social life in Musil's age can hardly be disputed—it was precisely this expanding role and status for which Musil reserved satirical intent in his portrait of "Kakania" in the first book of the major novel. It was also the pathological nature of this collectivity—the disease Musil felt his age to be suffering from—that led Musil to the parody alluded to in the title of book 2 and against which he cast his hero Ulrich and heroine Agathe, with studied irony, as "criminals."

## Transvaluing Values

Nietzsche's influence stands behind Musil's use of the idea of a "breach" or "break (-through)," an idea that is elevated to programmatic status in the title of book 2 of *The Man Without Qualities.* In this title, "Ins tausendjährige Reich (Die Verbrecher)" (Into the Thousand-Year Kingdom [The Criminals]), the word "Verbrecher" not only means someone who breaks the law, but also suggests a "breaking through" through to a new level of experience, a breach in tradition, even an overcoming of mores and established patterns of behavior. Nietzsche's "transvaluation of all values" from his major work *The Will to Power* (*Der Wille zur Macht*), I suggest, is a key influence on the scenarios and utopias developed in this part of the novel. These utopias are predicated on the requirement that the old must be cleared away before a new morality attuned to the life-enhancing needs of the individual becomes conceivable. Nietzsche demands an extreme negative stance, even extending to outright militancy, in order to eradicate all ossified forms of thinking. As he says: "Extreme positions are not supplanted by moderate positions, but, on the contrary, by other extreme, but opposite, ones."[10] As Ulrich and Agathe slowly feel their way toward the idyll of their "thousand-year kingdom" in the unfinished final section of the great novel, the purity of the demand for something new turns the couple into "Verbrecher," or, we might also say, code violators, destroyers of tradition. At other points Ulrich comes across as "outright antisocial and criminal" (*geradezu unsozial und verbrecherisch*; *GW* 2:722–23); for a significant period he concerns himself with the rapist and serial killer Moosbrugger, reflects on the nature of Moosbrugger's criminality, and campaigns for a new court hearing in order to clarify unanswered questions about the mental condition of the murderer and the significance his case possesses for the cause of justice in Kakania. In taking a stance on this set of thoughts as set out in his diaries, Musil initially attaches only legal significance to the notion of "criminality" (see *TB* 1:388). But Musil then extends the use of this term for his own purposes. These purposes are elucidated, for example, in Musil's critique of Schiller's notion of the moral theatrical ideal. Here the essential problem of classical theater is explained in terms of the inability of characters to transcend the morality imposed on them from outside (*TB* 1:479). While such a static notion of morality does invest humanity in the same manner as Christian principles of conduct with regular forms of thought and behavior, it can hardly do justice to the constantly mutable situation

[10] Nietzsche 1992, 52.

that individuals encounter in modern life, where, in Musil's words, life demands a "constant expansion" of the effort of the individual in order not "to be pressed down by the world" (*TB* 1:388).

This perspective, indeed, already implies a new way of approaching the challenges that life brings with it. Viewed from this position, life should no longer be taken to draw out from the individual a faithful compliance with particular moral commands, in order to simplify a complexly experienced individual existence, but rather puts that individual into a context of permanent movement requiring her to arrange the moments of correspondence with ethical demands on a case by case basis. The responsibility of people for their actions in the world is thus fundamentally their preparedness to shape their lives in the same dynamic ways in which the events of the world are revealed to occur. This, in turn, creates the conditions for a new understanding of ethical action that, in view of the diverse and not infrequently ambivalent circumstances encountered in life, require that the standards of received morality be departed from.

Musil sums up his ethical stance in the following way: "(The world flows in or one is streamed into the world) . . . moral imagination is created" (*TB* 1:388). To be emphasized in Musil's understanding of the term "criminal," for this reason, is the self-willed, creative dimension to morality that one finds, for example, in an unpublished variant of the chapter Musil was working on in the last years of his life, "Breaths of a Summer's Day" (*Atemzüge eines Sommertags*). In this draft version of the chapter, Ulrich, a "bold crosser of borders," obtains the secret password that permits him "to gain entry to the thousand-year kingdom or other ecstatic worlds of clear mindedness" (*GW* 4:1245). In this way, in Musil's imagining, the "Verbrecher" is no criminal at all, but an independent thinker who is ready to take his search for value into his own hands and thus break with the lawfulness of the moral code that otherwise binds him. Musil's apparent purpose with this idea of boundary-crossing and taboo-breaching is to reimagine the ethical. The vehicle for doing so in the final part of the novel is the emotional bond that develops between his paired protagonists, the "twin" siblings. In this way, and then subsequently with the creation of the sibling idyll, Musil aims to promote discussion of the circumstances in which new possibilities for life might arise.[11]

[11] Hartmut Böhme links this view of ethical literature to Musil's earlier work *Unions*; his goal in that work, as he puts it, is to expose the full compass of "human spaces of the imagination and possibilities for action . . ., in order to secure an unknown territory that opens up exemplary and not yet uncovered prospects for possible humanity" (Böhme 1993, 193).

## Musil's Ethics

In volume 25 of Musil's diaries, the product of his reflection in the years 1921–23, when he was working on the major novel, Musil makes several important statements about his concept of value. These statements hark back to a central distinction he draws between ethics and morality and between the ethicist and the moralist. Because of its significance for our discussion, this passage bears quotation at length:

> The moralist puts statements of a civil and moral (*sittlich*) nature inherited or taken over [from the past] into a logical order. He does not add value to these values so much as systematize them. Precepts, principles . . . are relative positions (*Lagen*) in a system. . . . The impulse (*Trieb*) that guides him is that of logic. He deploys ethical statements (*Sätze*) only insofar as they can be applied logically. All philosophers tend to view ethics as an attachment (*Anhängsel*) to theoretical philosophy. They have few ideas of their own . . .
>
> A special class of thinkers (*Forscher*) for whom moral civility (*Sittlichkeit*) is understood in terms of the relations that comprise it is worthy of attention. The names [that feature here]: Kung-fu-tse, Lao-tse, Christ and Christianity, Nietzsche, the mystics, the essayists. The Stoics and Epicureans. . . . Their contribution to ethics concerns not form but material.
>
> They undergo new ethical experiences.
>
> They are human beings of a different kind (*andere Menschen*). . . .
>
> They are the teachers of humanity.
>
> There is no single teaching (*Lehre*) about the human being. An ethics.
>
> The teaching [about the human being] is to be found in subjective conditions. . . . (*TB* 1:645)

The distinction that assumes importance for Musil's concept of value is therefore that between morality and ethics. Morality, in Musil's eyes, is an encrusted type of thinking that is meant to resist corrosion and change. Morality, as most philosophers of Musil's age would understand it, is the attempt to arrive at value statements prescriptively in a way that ignores the dynamic nature of the situations in which ethical thought is encountered. Against such morality is to be found something Musil calls "ethics," or, perhaps better, "the ethical." The ethical valorizes the circumstances that impinge on what Musil thinks of as the material of individual experience. This material is situatively experienced, mutable, and subjective. To experience the ethical is to have an individual experience of a particular moment in time from an intensely private perspective that is mostly not covered or described by the predictive conceits of moral precepts. The ethical typically lies outside the reach of such precepts. The thinking associated with these precepts and this morality, as Musil says, is "paralysing": "It shuts out the the individual. Which we sense as an ethical source. [Such morality] is dead" (*TB* 1:644).

In the Nietzschean atmosphere that Musil's works cultivate and which *The Man Without Qualities* sustains throughout, ethical decisions grow out of notionally indefinable, undelimitable, individual experiences that arise in human emotions in the domain of the non-ratioïd. This atmosphere contrasts with prevalent moral understanding in Musil's society, which assumes that morality is transacted predictively against agreed moral standards, or, in the Kantian view, where action is called upon to align with a super-principle—the categorical imperative—according to which that action would express, or be able to state, a universally valid moral law. In the sibling idyll around which the events of book 2 of *The Man Without Qualities* are constructed, Musil aims, despite the singularity his ethical thought demands, for a "certain type of objectivity" (*TB* 1:645). As a diary entry explains, this is an objectivity, noted down the ages by exceptional thinkers but largely missing from Musil's time, in regard to which humans have felt the pull of a particular form of melded thought and feeling arising from individual human experience and have attempted, in discussion with like-minded souls, to live according to it.

# 7: *Unions*: "An Episode of More than Merely Personal Significance"

THE YEARS FOLLOWING the release of Musil's first work, *The Confusions of Young Törless*, were of decisive significance for the budding writer. Towards the end of 1906 or in early 1907 Musil met his future wife, Martha Marcovaldi (1874–1949), six years his senior. In the following year he concluded his studies in philosophy and psychology at the Friedrich Wilhelm University in Berlin with a doctoral thesis on the physicist and philosopher Ernst Mach (discussed in detail in chapter 2). The thesis was given the grade of "laudable" and later was to bring about the offer of an assistant position with the noted professor of psychology in Graz, Alexius Meinong. This offer represented a chance to embark on a further academic project, qualify as a professor, and establish a university career. As attractive as this offer undoubtedly was to Musil, however, the unexpectedly auspicious debut of *The Confusions of Young Törless* had put him in mind of a literary career. This did not mean that he turned his back on the investigation of inner states that he had begun to address with such success in that first novel. He had no intention of abandoning this type of psychological project. Rather, he did not see the environment of the university, with its interest in articulating definitive general statements about the social behavior of human beings, as being conducive to the sort of project to which he was increasingly committed.

As Musil was later to remark in the foreword to his last published volume, *Posthumous Papers of a Living Author*, the project he was keen to develop further at this time involved the question of the relationship of individuals to their ideals (*GW* 7: 972). Clearly, any exclusively empirical approach to questions such as this would quickly hit upon limits. Musil's interest, by contrast, lay with the kind of complex value attributions and investments human actors make when the need to act presses upon them. These attributions could only hope to be partially elicited through external observation and conventional scientific inquiry. The academic approach to the question of human ideals seemed to Musil, for this reason, to fall short of the complex understanding of human motivation and behavior his project aimed for. Despite this, his intention was to embark on a credible intellectual path toward arguably the same results on which scientific analysis was premised. As Musil began to conceive this path in his early years, this was an approach that sought to place knowledge and artistic understanding, belief and mathematical reckoning, stark calculation and mystical love on an equal footing. The word "Vereinigungen" ("unifications" or "unions"), which was to become the title of the volume of stories he was envisaging, therefore sought to make evident ostensibly opposed types of knowledge and insight that would lead to an entirely new disclosure of the real. For this reason alone, as Musil observed in his diaries, this project carried with it from the beginning something "of more than merely personal significance" (*TB* 1:215).

Such a project, which sought alignment between domains usually kept apart under the academically scientific dispensation of his time, was promoted, in its early form, by a set of aspirations that would later grow into a life's work. It is the link between the novellas' raw and unformed material of a still-fledgling project seeking to forge unities between disparate realms of thought and the mature form of this project in Musil's later work, most notably, of course, the major novel *The Man Without Qualities*, that makes these novellas significant from our vantage point today.

## Biographical Factors

In the period after *The Confusions of Young Törless* the question of how human beings understand their own contradictory inner purposes was of deep personal significance for Musil. His friendship with Martha Marcovaldi was deepening: in August 1908 Musil became acquainted with Martha's two children. At the same time, disturbed relations between Martha and her husband had plagued the new couple for a prolonged period after Martha's separation from her husband and their divorce toward the end of 1907. These difficulties spilled over into the new relationship and became a source of tension, notwithstanding the fact that this relationship had quickly become warm and loving. The new literary project, which was commissioned by the editor Franz Blei at the end of 1907 or the beginning of 1908 and which ultimately grew into the two novellas of *Unions*, emerged from this mixed atmosphere of new love and sexual passion on one side and ongoing frustration and jealousy on the other. The pivotal moment when these tensions broke through to the surface, however, did not occur until an event came to light that Musil would later describe as a "catastrophe" (Corino 2005, 384): Martha confessed to having an affair with a long-time friend, Martin Cohn.

If Musil had revelled in the intense euphoria of passion in the early stages of his relationship with Martha, he soon saw himself confronted with the sober knowledge that loving a person and remaining faithful to that person might in actuality be two different things. It is likely that from Martha's infidelity and, as Corino speculates, additional repressed feelings of jealousy felt by Musil in relation to Martha's deceased first husband Fritz Alexander (Corino 2005, 383), for whom Martha appears to have felt profound love, Musil drew energies at an advanced stage of the writing process that quickly saw the project through to completion. At any rate, a piece that had been mired in periodic stasis for almost two years under the title *Das verzauberte Haus* (*The Enchanted House*) was converted relatively rapidly over just a few months into its final form under the new title *Die Versuchung der stillen Veronika* (*The Temptation of Quiet Veronika*). This sudden increase in the pace of work, which initially applied only to *The Temptation*, soon carried over to the last phase of work on *Die Vollendung der Liebe* (*The Completion of Love*), such that Musil was able to record the precise point in time at which he concluded both stories: 3:05 p.m. on November 19, 1910 (*The Completion*), and 1:00 a.m. on January 11, 1911 (*The Temptation*). The novellas were then put together under the title of *Vereinigungen* and published a few months later by the Munich publishing house Georg Müller.

The fastidiously precise notation regarding the completion of the work on *Unions* suggests that for Musil the novellas are not only meant as an investigation of the inner life of two women who pass through various stages of fascination or "enchantment" with the sexual (from the initial title of the story *The Temptation of Quiet Veronika*). It also suggests a meditation on Musil's own inner processes as he attempted to understand the power of the sexual drive to entrance the woman he was preparing to marry as well as himself. A second feature of *Unions* can also be considered personal: it concerns the path of writing on which Musil now found himself. As already suggested, Musil had consciously embraced a career of writing in hopes of gaining the intellectual freedom he believed would be denied him if he followed an academic career. In turning away from academia and committing to the profession of writing, Musil well understood that he would quickly need to establish settled domestic circumstances. For this reason the completion of work on *Unions* was followed by marriage to Martha just a few months later. Even before this, Musil's father had assisted this process of finding settled circumstances, helping his son to find a post in the library of a technical college in Vienna. One can say, therefore, that upon completion of this project Musil was able to celebrate a triple "Vereinigung" or "consummation" of his own interests: the triumph of his love for Martha, the consolidation of his career as a writer, and the establishment of settled financial circumstances on his assumption of paid employment as a librarian in Vienna.

This threefold consummation, and the goal of bourgeois solidity it implied, may at first seem at odds with the socially and morally progressive nature of the project as a whole. In each of the two love stories, after all, the aim of love is union with the intended beloved; the stories tell of the fact that libidinous stirrings within the protagonists intervene and derail the higher interests of this love. Nevertheless, the embedded conservative element, if assumed to be central, would not explain or account for other impulses in these narratives that seem directed toward emphasizing how labile the sexual impulse really is and how arbitrary the social and moral boundaries appear to be that are meant to corral it. One could even say that the stories realize a calculated provocation in this regard, since Claudine, the focalizing character of the story *The Completion of Love*, degrades herself with a strange man ("the word sodomy struck her; should I now perform an act of sodomy?," *GW* 6:180), and Veronika's sexual feelings in *The Temptation* are ignited by the sight of an aroused Saint Bernard dog. In light of the primacy of human freedom that the stories more or less insist on and hold to be non-negotiable for the human being, the overriding intention of the novellas appears to be nothing less than a fundamental reimagining of the entire region of human morality.[1]

The radical nature of this project is one reason Musil's public could make little sense of the stories after their appearance. Nor did the stories meet with critical success—*Unions* was savaged by the critics, and those who had been supportive of the earlier novel, such as the leading critic Alfred Kerr, this time chose to remain silent. It has since been noted[2] that the relationship between Musil's private interest, which turned on his relationship with Martha, and his public interest, which had to do with an independent literary program, were never effectively disambiguated in the project. It is only in the sibling idyll in the second section of the major novel nearly twenty years later that these elements came together in an effective manner—a private utopia on the one hand and a literary program seeking the establishment of a new type of morality and a "new humanity" on the other. Yet even in this more successfully combined later form, the project that *Unions* ushered in was to remain inconclusive: Musil did not finish the major novel and therefore was never able to give final form to the attempt to redraw the landscape of ethics.

## Intuition and Inner Life

The premise for the inquiry the novellas undertake is built on an unusual proposition about human emotions: that it is a set of emotional circumstances seemingly at odds with the ambition of love that actually realizes the goals of love on a different level. In the first story, a devoted,

[1] In a short paper from this time, Musil comments in this regard: "Actually, it is from the moral arena that one derives the prototypical sense of lawfulness. But it is now high time one eliminated this concept of lawfulness from morality" (*GW* 8:1306).

[2] See for example the critique of J. M. Coetzee of this aspect of *Vereinigungen*; J. M. Coetzee, "On the Edge of Revelation," http://jikje.home.xs4all.nl/Essay/coetzee.html (accessed January 20, 2012).

loving wife, despite her feelings for her husband, is moved to "betray" him with another man; through her betrayal of her husband she becomes acquainted with the depth of her love for him. In the second story, a bourgeois woman must transform herself into an animal if she wishes to become acquainted with the complete nature of the love she feels for her chosen lover. It is in such unusual circumstances that Musil seeks to locate the source of value attributions that—according to the project Musil follows—in turn can lay claim to constituting a source of ethical orientation for the human individual. How this link between raw sensual experience and ethical experience functions in practice lies at the heart of project the stories discuss and seek to clarify.

To gain additional purchase on the sort of intellectual terrain the stories traverse, it is valuable to consult the thought of one of Musil's noted contemporaries, Henri Bergson.[3] Bergson makes this point about the pitfalls of self-awareness in *The Creative Mind*:

> Everyone [will] have noticed that it is more difficult to make progress in the knowledge of oneself than in the knowledge of the external world. Outside oneself, the effort to learn is natural; one makes it with increasing facility; one applies rules. Within, attention must remain tense and progress becomes more and more painful; it is as though one were going against the natural bent. Is there not something surprising in this? We are internal to ourselves, and our personality is what we should know best. Yet such is not the case; our mind is as if it were in a strange land, whereas matter is familiar to it and in it the mind is at home. (Bergson 1992, 41)

These remarks introduce a problem that Musil had already touched on in his first novel, *The Confusions of Young Törless*—a problem now taken up as the explicit focus of *Unions*. The problem concerns two ways of gaining clarity in the world, or, as Bergson puts it, two manifestations of "the effort to learn." The first way—the effort to gain knowledge of the external world—consists in a process of learning to acquire and master rules. Bergson goes on: "Our action is exerted upon matter, and the farther the knowledge of matter has been pursued the more efficacious is the action" (41–42). The second effort to learn, by contrast, looks inside the self but is immediately confronted by the problem that knowledge is uncertain,

[3] In a late diary entry recording Bergson's death, Musil indicates awareness of parallels between Bergson's and his own project. But he also claims that he was "never able to read" Bergson on account of a disagreement with the concept of *durée créatrice* and the distinction Bergson draws between space and time. Nevertheless, the familiarity Musil displays with key terms in Bergson's philosophy and the fact that he considers it important in this diary entry to investigate Bergson more fully at a later date underscores the basic sympathy of their approaches, which my comments here elucidate (see *TB* 1:960).

that progress in acquiring it is painful, that this effort finds itself all of a sudden "in a strange land."

Viewed in this way, we immediately notice a difference between, on one side, the methods we follow and the facility we develop when we seek knowledge of the material world, and, on the other, the methods we need to follow in order to obtain knowledge of our own minds and ourselves. In the case of the former, as Bergson argues, knowledge for the moderns is based on a particular kind of experience—an experience focused on "nature" in the broader sense that nevertheless only embraces a part of reality. The moderns distinguished themselves, in other words, by the preference given to the knowledge of outer reality. Perhaps some day, as Bergson also suggests, ratifying this type of knowledge project on its own ground, this knowledge will have reached "the bottom of that part," or at least "approach it indefinitely" (43). Today that type of knowledge is called science, but it used to be called metaphysics and comprise half of the program of knowledge. The second type of knowledge, for which the term metaphysics in a restricted sense is now commonly applied, is equally tutored by the fact of experience, but this experience is the inner experience of the self, which is to say, the experience that self gathers as it observes its own "inner" progress in the world. The approach to this experience, this knowledge of one's mind, is not elicited—as the Romantics seemed to have hoped—through deepening insight into nature as such, but, as Bergson puts it, on assumptions that outer nature and inner nature are distinct from each other, through "the direct vision of the mind by the mind" (42), that is, through intuition. Intuition is an intelligence of the mind that uses "ideas as a conveyance" (42), but the knowledge sought will not be reducible to those ideas and, in any case, will have to avail itself of comparison and metaphor in order to make good what "scientific language" alone cannot express.

Musil's outlook in *Unions* very closely approximates that which Bergson specifies in these framing statements from *The Creative Mind*. Musil's project is metaphysical in its focus on the contents of experience. Its objects are not principally those of outer experience—the objects cultivated by science. Rather, the project's main concern is the shape of inner experience as it comes to be felt over time by a specific human individual. In this domain of the self, time is of great significance. Bergson uses the term "durée" to stand for the lived sense of time that shapes inner experience and is fundamentally dissimilar to that external, chronological measure of time in which the objects of outer nature are observed to unfold. Like Bergson, Musil aims to establish the significance of this internal experience of time for the individual, even as he will later disagree with Bergson in his diaries about what this inner experience of time amounts to. To put the matter plainly and in a way that holds for both Bergson and Musil: we do not reliably know what the objects of inner experience truly are because, as moderns, we have chosen to be influenced more by matter than by mind, and thus we have also chosen to reverse the older dispensation that, for example in the Platonic theory of ideas, had privileged the latter (mind) over the former (matter). Musil's project is in some ways similar to the Platonic approach, but exhibits far less certainty about the status of the reason it cultivated and the confluence of the true, the good, and the beautiful it believed in—a confluence that organized the appearance of those objects for the mind that imagined them. For Musil—whose view of these matters, like Bergson's, is far more modest than that of the ancients—the status of reason in inner experience cannot be assumed in advance. The question of the comportment of truth with its partner ideals of ethics and aesthetics, equally, is something that can only be known more fully once inner experience has been investigated and, as far as possible, brought to a state of knowledge. In connecting with an older concept of experience already known to the ancients, Musil developed a metaphysical style of reasoning, which, however, steers its course without the rational support arguments that had structured ancient metaphysics and which, on the model of early mathematics, had given this older metaphysics a consistent and rule-bound appearance. It is the absence of these support arguments even more than the return to an older metaphysical approach to knowing that brought about the difficulties that dogged the early reception of Musil's stories.

For Musil, the support for the metaphysical inquiry he is interested in was to come in large part from the application of scientific method. To this extent, the disclosure of inner states was to be approached much like the objects of science itself: feelings presented as a field of emotions to be observed dispassionately; the "scientific gaze" would not only be directed toward the observation of inner states but also to the question of what changes in the pattern of these states are observable over time (a strict separation between the lived time of the subjects of observation and the external time of scientific observation itself would of course be preserved); the aim of the analysis was to identify shifts in the emotions that in turn would lead to the presumption of a broader change in emotional quality. It is this change in quality in the emotional field that would lead to the identification of a shift of emotional content onto a higher plane of the emotions—a change that would effect a movement in the emotional field that could be called ethical. In Musil's view, it was this higher-order change in the emotional field that constituted the ground, if not also the purpose, of ethical understanding itself.

To conclude this adumbration of the assumptions and intellectual background of the stories, Bergson is again helpful in shedding light on Musil's attempts to recast the boundaries between science and metaphysics. In the traditional view, as Bergson points out, science condemns metaphysics to "a hypothetical and vague knowledge," since science "will necessarily have taken to itself in advance everything precise and certain that can be known of its object" (43–44). For Bergson as for Musil, this suggests that only science can aim at precision, and indeed, the more it does so, the more metaphysics will appear imprecise, partial, and indeed unnatural. Yet as Bergson contends, and which equally holds for Musil, this division of labor is prejudiced:

> Quite different is the relation I establish between metaphysics and science. It is my belief that they are, or that they can become, equally precise and certain. They both bear upon reality itself. But each one of them retains only half of it so that one might see in them, if one wished, two subdivisions of science or two departments of metaphysics, if they did not mark divergent directions of the activity of thought. (44)

When we look closely at the early work *Unions*, then, we see that it continues Musil's interest in his dissertation in the question of science. The question of science, for Musil, is less a question of the objects of science and more a question of how science views these objects. For Musil no less than Bergson, the legitimate objects of science need not be restricted to outer reality but can just as easily be found in the inner life of human beings, where they stand the same chance through the exercise of scientific method of being intelligible. The two stories of *Unions*, from this angle, are not narratives at all so much as intellectual experiments designed to elicit the shape and characteristics of the inner reality of the human being. It is this experimental viewpoint that allows the stories to be seen as attempts to think scientifically about this inner reality and thus arguably to carry out a type of scientific experiment.

In this instance, the aim of the scientific experiment is to gain insight into the emotional responsiveness exhibited by the experimenter. Such responsiveness is cued by a complex of aspirations, sufferings, and physical urges that present as psychic entanglements and confusions in the emotional field. The scientific interest of the stories is the identification of resources by which to generate accurate understanding of these confusions and potentially to dispel them. In Musil's view, the transparency to emerge from this process will not, as with the pretension of Freud's psychoanalysis,[4] deliver conventional scientific content—say, in the form of testable general statements applicable to humanity as a whole. At most, Musil's experiments will result in higher respect for the function and

[4] Most of Musil's references to Freud in his diaries indicate disagreement with him: for example, when Freud, in Musil's view, falsely construes the desire of the child for closeness to its mother and warmth from her as exclusively sexual (*TB* 1:1011), or when he declares that Freud develops major insights, but mixes these with prejudice and dilettantism (*TB* 1:74).

value of the inner life and a description of the factors that, in specific instances, constitute the source of ethical understanding for a particular human being.

## Soul Experiments

As the foregoing has established, *Unions* proceeds from the assumption that there are two ways to approach experience, both of which exercise a valid right to be heard. Whereas one approach concerns itself with experience in relation to an external world, where the idea of nature expresses key elements of that experience, the other level is focused on the data of inner experience, where the main informant is the agency of inner intuition. Musil's conviction about this second level of experience, and the methodological provocation he presents to his readers, is the idea that this second region of experience can be understood, at least in theory, with as much precision and accuracy as the first. The impediment to precision lies principally with the fact that the language with which to chart the forms and characteristics of this experience is still embryonic. We see the attributes of the new language Musil seeks to apply to inner experience in the following passage:

> Somewhere one must be able to hear two voices. Perhaps they lie as if dumbly next to and within each other on the leaves of a diary, the dark, deep voice that is suddenly all about a woman, as the pages would have it, embraced by the soft, wide, extended voice of a man, by this entwining, incomplete voice, between which something she didn't have time to conceal looks out. Or perhaps it was not this at all. Perhaps there is a point somewhere in the world into which these two voices, only slightly jutting out from the tired confusion of daily voices, rush forward like two beams of light and become entangled, somewhere. Perhaps one should want to find this point . . .(*GW* 6:194)[5]

This passage introduces *The Temptation of Quiet Veronika*, the second story of the *Unions* project. If the language it announces comes across

[5] "Irgendwo muß man zwei Stimmen hören. Vielleicht liegen sie bloß wie stumm auf den Blättern eines Tagebuchs nebeneinander und ineinander, die dunkle, tiefe, plötzlich mit einem Sprung um sich selbst gestellte Stimme der Frau, wie die Seiten es fügen, von der weichen, weiten, gedehnten Stimme des Mannes umschlossen, von dieser verästelt, unfertig liegen gebliebenen Stimme, zwischen der das, was sie noch nicht zu bedecken Zeit fand, hervorschaut. Vielleicht auch dies nicht. Vielleicht aber gibt es irgendwo in der Welt einen Punkt, wohin diese zwei, überall sonst aus der matten Verwirrung der alltäglichen Geräusche sich kaum heraushebenden Stimmen wie zwei Strahlen schießen und sich ineinander schlingen, irgendwo, vielleicht sollte man diesen Punkt suchen wollen. . . ."

as neo-Romantic, a descendent, say of Novalis's long prose-poem *Heinrich von Ofterdingen*, this is because Romanticism, in some of its experimental expressions, comes closest to furnishing Musil with the type of approach he requires in order to carry out his project. Yet it would be a mistake to see the invention of a new language, a notable aspiration of the Romantics, as the key to this project. Rather, Musil's intention is to realize through a necessarily intuitive, and even idiosyncratic, grammar of inner experience a set of results that will in the end be conceptually, not linguistically, enabled and will seek to make an intellectual contribution to our knowledge of concepts. So even in an opening passage that cues the fact of inner experience through a new type of language, it is not the fact of this language that is noteworthy and for which Musil seeks to garner attention. Rather, Musil investigates the spaces that open up between the encrusted states of being that his two main characters in this story inhabit and from which their conceptual orientation in the world emerges. The world of inner experience is to be found in such spaces. The language of this experience is symbolic,[6] not propositional, emergent, not fully expressible, and it is pushed out into the open by the psychosexual impulse that drives the human being to seek attachments with other human beings. This inner experience remains hidden from the subject that experiences it; for this reason, as Bergson observes, it is "difficult to make progress" (Bergson 41). Yet it is also the case, though the solutions proposed will be incomplete and the conclusions only provisional, that "an increasing probability which can ultimately become the equivalent of certitude" can be achieved (46).

Both stories are arranged around strategic transgressions. For Claudine, in *The Completion of Love*, the transgression takes the form of the "obscure, incomprehensible appeal of a lie" (*GW* 6:187); for Veronika in *The Temptation of Quiet Veronika*, it is the "indescribably numbing descent into the life of the animal" (unsagbar gleichgültige[n] Herabgleiten des Tiers, *GW* 6:197). These are moments when the inherited moral dispensation of the bourgeois subject is thrown over and a space is cleared for something new to appear in its place. In the intimate recesses of individual experience where the physical and sexual response to environment comes to the fore, a redrafting of the terms of the ethical for these actors can be undertaken. It is to be noted that this view of the ethical contradicts the Aristotelian view that it is in social life, the life of the collective, that the instinct to make ethics ultimately issues.[7] Musil might agree that ethics is made from attachments relating to other people, but

[6] It is worth noting in this context the interest Musil later gives in his diaries to Cassirer's *Philosophy of Symbolic Forms* (*TB* 1:721).

[7] Cf. MacIntyre 1966, 100: "In Greek society the focus of moral life was the city-state; in the Hellenistic kingdoms and the Roman empire the sharp antithesis between the individual and the state is inescapable."

these concern not the social group (where, in the Aristotelian view, ethics is a cousin of the political), but the physical urge to connect with another human being in a psycho-sexual sense. In Musil's account, this physical urge occurs below the level of conscious calculation, where the emotions stir in profound, pre-rational ways. It is from this raw emotional material that ethics, as a deeply personal associative instinct with conceptual accretions, first emerges.

The moments in which an emotion of this sort takes shape can thus appear highly idiosyncratic. In *The Temptation*, the protagonist Veronika's affections, vacillating between the two male figures in the story, must first be confronted in the absence of one with the thrilling physical proximity of the other before there can be a change in the comportment of these emotions. When this change occurs, it brings about a change in the underlying direction of the narrative. The change is foreshadowed in reference to the Saint Bernard dog as a primitivization of emotions that ultimately enables Veronika to become acquainted with her own "true" inner "picture," her "*veraikon.*"[8] Musil makes such a "de-bourgeoisification" of her feelings, which is linked in the story with her physical arousal in the presence of Demeter, one of the two male figures in the story, a precondition for the transformation that takes hold of Veronika's whole person, finally bringing about an inner union with the absent Johannes. In *The Completion of Love*, a similar path of emotional transformation is illuminated. Claudine, the focal character of the story, taking the train to see her young daughter, experiences various phases of emotional closeness to, and distance from, her husband. The narrative centers on the problem of this waxing and waning of her emotions, which Claudine initially understands to signify an estrangement from her husband and her marriage. As the story reveals, however, a deep and direct experience of this emotional flux in the end provides the key to establishing the true shape of her feelings for her husband. On her journey Claudine meets a stranger who on one level means nothing to her. Yet, as the story progresses, exposure to this man and the prospect of intimacy with him eventually brings about a resolution on the plane of her emotions: while staying in a hotel, Claudine wants not only to go and see this man but also to debase herself with him sexually. It is the experience of primitive emotion in the presence of this man that finally triggers the change in her feelings in line with the ethical project that both stories follow:

> At this moment she felt that her body, which enveloped everything it sensed as if these things were familiar, had become an unclear barrier. She felt the sense of this man, which was encased around her more tightly than everything else, all of a sudden as an inescapable

[8] Cf Hartmut Böhme's interpretation in the epilogue of the Suhrkamp edition of *Vereinigungen* (1990, 210).

> betrayal that separated her from her beloved and, in an experience of powerlessness she had never known before, that sank down upon her, as if the last faith she felt in her body turned, on an unfathomable inner ground, into its opposite. (*GW* 6:186)[9]

The path of inner feelings traced in both stories is thus a path out of confusion and uncertainty into the transparency of an underlying emotion for which Musil wishes to make an ethical claim. This ethical claim is cued, in turn, by a higher voluntaristic state where an ethical alignment with another human being is at stake. This unusual "take" on the question of ethics—that the source of ethics is feeling-based rather than rational, and that ethics is as much a question of personal identity as a feeling relating to a social group or collective—seems at odds with the entire tradition of thought about ethics that has gone before. Musil justifies this rejection of tradition on the grounds that emotional confusion, the "deep-seated uncertainty of spiritual values" (tiefe Unsicherheit der seelischen Werte, *GW* 6:186), cannot be alleviated on its own by an intellectual orientation—such as moral duty, in the Kantian account. The path Musil advocates in *Unions*, instead, does not lead from emotion to a moral precept derived from the order of reason, but rather from one emotion to another significant emotion. On this unusual path, the ethical state is reached only when a succession of emotions is resolved in the time of inner experience in such a way that, under the rising pressure of psychic events, it is "completed" and reaches its final form. In *Unions* at least, this is an investigation of inner states that Musil means to approach scientifically.

[9] "In diesem Augenblick empfand sie ihren Körper, der alles, was er fühlte, wie eine Heimat umhegte, als eine unklare Hemmung. Sie spürte sein Gefühl von sich, das, näher als alles andere, um sie geschlossen war, mit einemmal wie eine unentrinnbare Treulosigkeit, die sie von dem Geliebten trennte, und in einem ohnmächtig auf sie niedersinkenden, noch nie gekannten Erlebnis war ihr, als verkehrte sich ihr die letzte Treue—die sie mit ihrem Körper wahrte—in einem unheimlichen innersten Grunde in ihr Widerspiel."

# 8: The Problem of Trust in *The Man Without Qualities*

MODERN LIFE in technologically advanced societies involves a considerable measure of complexity. How such complexity is to be borne by individuals appears to be an increasing problem for such societies.[1] In Niklas Luhmann's view, complexity is principally dealt with in these kinds of societies on the basis of the concept of trust. In Luhmann's analysis, social trust is an abstract concept whose aim is the conservation and promotion of intersubjective tolerance. Social trust in social groups works by analogy with trust between individuals, where a psychological gain is associated with the rewards that flow from successful trust strategies. Luhmann's approach sets out a comprehensive theoretical framework by which systems, environment, function, and complexity are understood to be structural components of a social theory of communication. According to Luhmann's understanding, society itself is already a form of communication and, in the reverse, communication a form of society: "Society cannot be conceived without communication, and communication, for its part, not without society. . . . The reproduction of communications by way of communications takes place in society" (Luhmann 1997, 13). As soon as social communication is approached in this matter, it is a short route to the conclusion that modern societies are faced with the real-world problem of complexity management. Luhmann says that social communication "contains more possibilities than that which the system can conserve reactively." Although a system adjusts selectively and autonomously to its environment, it is the human agents in the system who, in their awareness of the world's complexity and through the nature of their personal experience of their "world," face the referential problem of how to attend to their own individual self-conservation within the system.

With these starting assumptions in mind, I would now like to drill down into this problem of complexity in the context of Musil's novel *The Man Without Qualities.* In doing so, I touch upon a set of concerns that have been enlarged upon in literature by several authors over the

[1] Luhmann 2000, 19. In recent sociological research, the question of engagement with complexity has been intensively investigated using the guiding concept of "social capital." For examples of such studies, see Lin et al. (2001), Gittell and Vidal (1998).

last 200 years—one thinks, perhaps above all else in German literature, of the early-nineteenth century Prussian author Heinrich von Kleist (1788–1811), in whose works—as was recognized early in research on Kleist[2]—the question of trust appears front and center among his themes. While for Kleist, the difficulties of successful communication and the problems that occur when trust is compromised are framed as a problem of subjective orientation in a post-Enlightenment world, leading to assumptions about what manner of "hermeneutic orientation" might be required to deal with this issue of trust more adequately, for Musil the question of trust is recast as the problem of responding to an entirely new phenomenon in the evolution of human societies: the appearance of modern technocratic consciousness. Musil probes this new phenomenon in various works, but nowhere with more intensity or in greater depth than in the major novel. In this context, thinking in a "technical" mode with access to resources made available by technology affects the question of social trust in significant ways.

My aim in this chapter is to highlight the advantages of a systems theoretical approach to clarifying the problem of trust in *The Man Without Qualities.* This approach will free up new perspectives in relation to the analysis of society the novel puts forward, particularly in book 1 of the novel. My first task in what follows is to outline key features of the systems theoretical approach to the problem of trust as they have been expressed by Luhmann in two important writings: his early essay *Vertrauen: Ein Mechanismus der Reduktion sozialer Komplexität* (translated as: *Trust and Power*, Chichester: Wiley, 1979) and his later, major work *Die Gesellschaft der Gesellschaft* (translated as: *Theory of Society*, Stanford: Stanford University Press, 2012).

## Luhmann's Social Analysis

In Luhmann's approach, the world-concept of the individual does not in fact create a "world" at all, but rather at most only a personal "system structure" and "foundation for behavior." This foundation is enhanced by recognizing the world-concepts of others as an expression of "egocentricities" understood by the individual to be no less valid than her own. This augmenting of the individual's experiential world through openness toward the concepts of others then plays out at a social level. For Luhmann, the fact of increasing complexity signaled by these acts of augmentation functions as an "attractor" for the evolution of society as a whole. At the same time, increasing complexity is bought at the price of higher levels of general social "conditioning" (Luhmann 1997, 741). The intersubjective sphere of sociality thereby becomes the carrier for a general

[2] See, for example, Fricke (1929) and Müller-Seidel (1961).

level of complexity that both projects emotional uncertainty onto the ego and inevitably increases the psychological burden felt at the level of the ego. Luhmann concludes from these circumstances that the individual is called upon to develop more effective forms of complexity management (Luhmann 2000, 8). In practical terms, such management of complexity is governed by mechanisms that function to standardize the experiential possibilities of individuals. These mechanisms both stabilize and delimit the capacity of these individuals to influence their environment while simultaneously preserving and expanding the complexity-potential of the social system in which they take place at a macro level. In this way, "expansion and reduction belong together as complementary aspects of the structure of human behavior" (ibid.). These mechanisms have the dual task of keeping the complexity-factors of the social system at a tolerable level from the psychological viewpoint of the individual, while simultaneously anticipating aspects of the future so that further complexity can be negotiated and integrated into the system. From this standpoint, both the reduction and the expansion of complexity describe core aspects of social and psychological risk management in the highly developed social systems of the modern era.

The factor of trust is one of the key elements on which the task of complexity reduction in the social system devolves. According to Luhmann, "where there is trust, there are more possibilities for experience and action, and the complexity of the social system increases, that is, the number of possibilities that can be accommodated through its structure increases" (ibid. 8–9). Since the function of a system is to differentiate its internal relations from the environment external to it, the dependence of a system on the mechanisms of trust generally increases over time, rather than decreases. Moreover, trust as a functionality stands in contrast to instrumental methods of control that equally seek to influence the future and are also aligned with the goal of complexity reduction. The particular function that is assigned to trust is defined not as intolerance for growing diversity, a feature of the instrumental exercise of power, but rather as a strategy of general "tolerance for ambiguity" (ibid. 19). Trust makes the lived experience of individuals more robust, sensitizes people to further acceptance of complexity, and enables them "to live and act with greater complexity in relation to events" that occur in the world (ibid.).

For Luhmann, this functionalist access to the problem of trust in no way takes place outside historical time. Luhmann considers an especially important historical event underlying the problem of trust to be the Cartesian revolution in the history of metaphysics in the seventeenth century, according to which "the consciousness of a thought thinking itself was put in place of [the notion] of the being of an existing [thing] for which consciousness had been laid before it" (ibid. 25).[3] This focus on a new type of conceptual awareness that reflected the terms of thought, rather than a directly experienced external world, back onto itself, was the grounding moment of a new knowledge project. But it also had the consequence of projecting the issue of self-consciousness onto the inner life of individuals. As modern societies arose, societies erected on an Cartesian epistemology, under whose terms the issue of self-consciousness became a private question of accommodation to external complexity, the need for effective forms of complexity reduction was felt as an increasingly strident demand.

In this respect, the position of the positive sciences—the enablers of new and increasingly "technical" knowledge—contrasted with the psychological situation of individuals on the cusp of the modern era. According to Luhmann, in line with this Cartesian development, "truth-effective truth" (*wahrheitsfähige Wahrheit*) increasingly became a question of correspondence between percept and concept for the positive sciences. The question of intersubjective certainty, and "whether it was sensible to substitute this question of certainty for the long trusted criterion of evidence as a truth category" (ibid.), has remained unanswered. For the real problem of intersubjective certainty, which is a feature of advanced societies built on the Cartesian model, the positive sciences offer a weak kind of consoling knowledge about the world—a fact Musil must have understood at least from the moment of his decision to turn his back on an academic career to become a writer, that is, around 1909–10. From this time, it seems that it became increasingly clear to Musil that the kind of regularity resulting from knowledge increase in the positive sciences on Cartesian (and later also Kantian) assumptions does not rest on the discovery of objective or "real" principles, but gains legitimacy rather from *statistical measures.* The role that statistics plays in ordinary life is one of the chief themes, for example, of Musil's drama *The Enthuasiasts* (1921). In that work, the investigations of the private investigator Stader are predicated on the view that statistics provides an means of comprehending the behavior of individuals, not just of the group. But, as events in *The Enthusiasts* makes clear, individual behavior cannot be approached with this kind of social calculus in mind. For Musil, the use of statistics as a social measure at most has utility for understanding group behavior. At that level, the greater the number of individuals in a statistical sample or "data set," the greater the statistical success overall. Statistics and the science that relies on this new field, therefore, cannot be expected to provide useful information about the voluntarism of ordinary human beings.

[3] Luhmann's original German is as follows: "die Bewußtheit des an sich selbst denkenden Denkens an jene Stelle, die vordem das dem Bewußtsein vorliegende Sein des Seienden eingenommen [hatte]."

This insight leads Musil to the kind of conceits that are deployed in the social discussion of Kakania in the later major novel and the picture of individuality that is cast against it.

*The Man Without Qualities* thus engages from the outset with the success or otherwise flowing from the statistical modeling of ordinary behavior. Indeed, a "man without qualities" is a label that underscores the fact that the wider social system of which he is a part has rendered the individual's experience of life to a large extent superfluous. As Musil notes in a diary entry, however, the statistical sciences are not in a position to provide individuals with their life goals (cf. *TB* 1:522), and so a gap opens up between the upscaling tendency of statistical measures and the downscaling interests of individual life. Against the backdrop of these considerations, Robert Musil's novel *The Man Without Qualities* launches an inquiry into precisely what it is—or should be—that constitutes these individual interests.

## Systems and the Individual

The issue of the comportment of the individual in modern life toward the fact and problem of rising social complexity is formulated in the novel as the question of the "right way to live" (*GW* 1:255). Musil approaches this question not in the manner of traditional narrative from the perspective of the experience of focalizing characters in a story about these characters, but rather from a standpoint that highlights the functionalist viewpoint. To cue this narrative move, the introductory chapter begins with a discussion of the abstract phenomenon of weather, in which ordered conditions, such as those expressed by the phrase "a beautiful August day," arise only fleetingly and then vanish again. That this real world, seen at this level of abstraction, does not in fact tend towards order, but rather must be imagined as subject to a continual process of entropy in the sense of Erwin Schrödinger's laws of thermodynamics as these apply to open systems (see *TB* 1:524), can be seen from the traffic accident in the first chapter, which is understood as a technically induced disruption in the rhythms of two subsystems, one human, the other machine. In this way the novel frames the issue it focuses on as the question of the "right" relation between the human and the technical. The commentary of the unnamed figure—the first direct dialog in the novel—immediately outlines a demand stemming from this relation between the human and the technical, namely the need to restore psychic order: "[These] heavy lorries as they are used here have an overly long braking distance" (*GW* 1:11). Shortly after, this figure quotes statistics pertaining to the number of deaths and injuries in car accidents in the United States. Thus, already in the first chapter of *The Man Without Qualities*, the functionalist perspective affords the insight that the social systems constructed by human beings in the modern era may not be able, or at least not reliably able, to reduce the complexity to which modern technologized human life is now subjected at the individual level.[4] These circumstances, in turn, visit a psychic need for order on the individual—a need in regard to which the reference to statistical understanding represents a psychological response. Despite the need for containment of the psychological risks facing individuals outlined in the first chapter in the novel, the risks attending the relation between the human and the technical seem undiminished and ever-present.

The novel underscores the fact that individual actors constantly face significant psychological challenges in the new technical spaces they inhabit in modernity. Any deviation from the general attitude, what Musil calls the general "direction of life" (*GW* 1:128), entails risk. Even the act of walking might be considered risky. As the narrator explains—not without Musil's signature irony—if one displays the slightest timidity, or the vaguest astonishment, about this act of "falling forward into the future," a disruption in the general state of mind can result and "one is no longer able to stand upright" (ibid.). The effort of effecting forward movement along ordered paths thus absorbs a large portion of the energy individuals expend every day. As Ulrich puts it, the level of attentiveness an individual needs to apply to the task of daily living would probably exceed that which Atlas needed to lift up the world (*GW* 1:12). It is therefore quite understandable if the execution of the daily tasks of life absorbs the concentration of individual people in such a way "that the life they lead and that leads them does not concern them inwardly much at all" (*GW* 1:129).

For Musil, the problem of psychologically overloading individual people is therefore one of the most significant factors of living in modern society. How this overload is to be managed under conditions of increasing complexity in the Austro-Hungarian society of the turn of the twentieth century, and how the life possibilities of individual people can be safeguarded and even enhanced under such conditions, becomes a central question in the novel. The answer to this question does not lie in a strategy of optimal complexity reduction alone. The aspect of complexity expansion, which is also a requisite feature of modernity, simultaneously represents a challenge as well as an opportunity for individual actors. In this context a notable shift in emphasis is perceptible in Musil's novel. Although the individual person, represented by the protagonist Ulrich, increasingly concerns himself with the question of complexity *expansion*, in order to find a more meaningful (or the "right") feeling for life,

[4] The entropic movement indicated in the first chapter of *Der Mann ohne Eigenschaften* was also conspicuous for Dahan-Gaida (1993/4) and Krommer and Kümmel (1993/4). Two important studies have related the entropic movement of the novel to Heisenberg's concept of complementarity (Kochs 1996) and Friedrich Kittler's theory of information (Kümmel 2001).

the social system in which this exemplary search for meaning takes place appears to be focused solely on a system-wide struggle *against* complexity. As such, both aspects of dealing with complexity appear in the novel not as complementary strategies, as described by Luhmann, but rather as *opposing* ones, which evince ever fewer commonalities, where each strategy increasingly puts the other into question.

The need to reduce complexity is therefore self-evidently a major concern of Austro-Hungarian society. The ironic name "Kakania," in this way, does not just stand for the spiralling complexity of a multiethnic state with its intricate and highly evolved government and administrative arrangements, but also holds, perhaps more pointedly, for its simplest operations: "According to its constitution it was to be governed in a liberal way, but in fact it was governed clerically. There was clerical rule, but one lived freely and independently (*lebte freisinnig*)" (*GW* 1:33). Underlying the kind of freedom practiced in this state, nevertheless, is a general restriction, because "one was negatively free in [this state]; one was continually surrounded by a feeling of the insufficient grounds for one's own existence and by the great sense of things that didn't take place" (*GW* 1:35). The "most advanced state" (ibid.) in the world at the turn of the twentieth century is thus not defined by a variety and multiplicity of options and choices, but rather by a scarcity of them, indeed by a certain "eventlessness," in which it is precisely non-happenings that emerge as typical features of the system. That these non-occurrences are not traceable to manipulation at the macro level, but rather appear as a reaction to the high complexity to which the system itself is generally exposed, is highlighted by the label "semblances occur" (*Seinesgleichen geschieht*), the title of the second part of book 1 of the novel. The label "semblances occur" emphasizes that the social processes occurring in the system have the primary function of proscribing diversity and thereby restricting what can be experienced at the individual level. What this proscription gives rise to, however, is the sense that events do not in fact take place. The psychological advantage accruing to this pre-patterning of the event horizon of individuals is that diversity can be managed in advance and absorbed into the system in a non-threatening way. In consequence, it is not the event-character of occurrences that comes to the fore, but rather the non-occurrence of a certain eventfulness—that aspect of the event that, although it is external and "new" to the system, is unable to effect any change in the system as a whole. Musil refers to this tendency of the Kakanian system to disable novelty and disruptive externality under the "principle of insufficient reason" (*GW* 1:133–37), a principle that turns the Leibnizian argument about rational action on its head. An essential trait of the Kakanian system is therefore a concern on the inside of the system to combat precisely the intrusion of the external complexity that would otherwise overload the system. But this also means that every action now lacks the rational motivation that, in the old view, created the grounds for all action to occur.[5]

In the second part of book 1 Musil then examines how this strategy of complexity reduction in the Kakanian system looks in detail. It is Graf Leinsdorf who gives the strategy its clearest expression: "It was clear to him [Leinsdorf] that something had to happen that would put Austria above every other state, so that 'this glorious promulgation of the life of Austria' would become a mark of distinction for the entire world, enabling her to discover once more her true inner being" (*GW* 1:88). An advantage of this endeavor to bring about this "glorious promulgation" of the Kakanian state is that nothing fundamentally new need occur. From this angle, the seventieth anniversary of the Austrian monarch, which becomes the focus of the campaign of the "parallel action," can be seen not as an attempt systemically to expand social complexity through an international celebration of the Austrian state, but rather the opposite: a duplication of an anniversary already initiated in the neighboring German state that is directed at meeting and disarming the external complexity that impinges upon the Kakanian system. Precisely therein lies the paradoxical form of the parallel campaign: while ostensibly open to a stimulus flowing from a set of external circumstances and a projected future occurrence (an imperial celebration occurring in another European state five years hence), the parallel action only demonstrates a concern to confirm the simplicity enshrined in the inner structure of the Kakanian system. Behind this campaign, however, and in a sense obscured by it, lies the informational fatigue felt in a general way by the ordinary denizens of Kakania.

The informational overload of ordinary citizens is then thematized in a passage which links the "suffering of contemporary people" to the type of "civilization" that has been fashioned for them: "[Civilization] is an obstructive condition, full of soap, wireless waves, the pretentious sign language of mathematical and chemical formulae, national economy, experimental research and the incapacity to establish a simple, but elevated community of human beings" (*GW* 1:103).[6] If civilization has brought on such rampant complexity, civilization can also heal it by effecting a unity of the commitments of this age with the soul (ibid.). This is how the educated grand dame Diotima, who quickly becomes the intellectual and local center of the parallel campaign, understands the aspirations of the parallel action whose meetings occur in her home. While "civilization was . . .

[5] For an introductory discussion of this inversion of Leibnizian thinking, see Mehigan 2001, 74–79.

[6] In the original: "Es ist ein hinderlicher Zustand, voll von Seife, drahtlosen Wellen, der anmaßenden Zeichensprache mathematischer und chemischer Formeln, Nationalökonomie, experimenteller Forschung und der Unfähigkeit zu einem einfachen, aber gehobenen Beisammensein der Menschen" (*GW* 1:103).

everything her intellect could not control," the return of control would be assayed through the resuscitation of the lost "elevated community of human beings" (ibid.). At the level of the social system, the coded form of this psychic desire for control takes more or less the same form. But the ambition to copy a piece of world-reality and paste it onto its own projections of the future—the ambition of the parallel action—in fact reveals the existence of an inner crisis of systemic trust within the Kakanian state.

This crisis of trust expresses itself in the sinking feeling of the individual that the individual is no longer capable of living truthfully. For the protagonist, Ulrich, there develops, in consequence, "the nagging suspicion that in this world false, careless, and personally unimportant statements echo with more force than the most intimate and real ones" (*GW* 1:129). This exchange of truth and falsity, of throwaway impersonal utterances and privately significant and meaningful ones, indeed the tendency to put higher informational value on statements that serve the group, even if they are valueless or false, than any truth that serves the needs of individuals, may be taken as a significant aspect of the crisis of the Kakanian state in itself. On the other hand, it is precisely in this stereotypical group behavior that we find the roots of de-individuation that in fact weakens trust in the system's structures and consequently destabilizes the system as a whole. Musil criticizes the merely technical, outward quality of this correspondence between internal experience (at the level of the subject's perception) and external experience (at the object level of the system) on the grounds that it appears to completely break with the unsystematizable aspects of life that the individual experiences as genuine and even unique. Musil compares this eclipse of individual consciousness with a "mold into which the contents flows like gas into a glass balloon," thereby generating "a technical way of being" (*eine Technik des Seins*, *GW*: 131). Furthermore, it is this technologized being that gives individuals for whom their own individual experience of life is vital the sense that they are no longer authentically alive. As the narrator then puts it, the ego is turned into an "external shell," even a "pseudo-ego," into which a "roughly approximate group-soul is inserted," and "if one pays a modicum of attention, one can always recognize in the newly arrived, most recent version of the future the good old days just around the corner" (*GW* 1:132).[7]

## Trust and the Problem of Trust-Worlds

As will have become clear, Musil's novel shines a light on a systemic feature of the Kakanian social system. This is its habit of imposing

---

[7] Musil's original German: "wenn man bloß ein bißchen achtgibt, kann man wohl immer in der soeben eingetroffenen letzten Zukunft schon die kommende Alte Zeit sehen" (*GW* 1:132).

prefabricated modes of behavior on individuals—a mechanism of complexity reduction by which the aspect of free choice at the system level is forfeited or neutralized, and where choice is a selection that becomes inseparable from the category of familiarity. Musil's societal critique suggests that the stabilizing aspect of "familiarity" (*Vertrautheit*), which Luhmann sees as a "prerequisite for trust" (Luhmann 1968, 22–23), has itself become a problem. In trust-worlds, according to Luhmann, the past dominates over the present and future: "In the past there are no longer 'other possibilities,' the past is always already reduced complexity" (Luhmann 1968, 23). If, however, the anticipation of the future gives back no new elements of life experience, but rather only more of the already experienced past, the purpose of the passage of time, upon which experiential possibilities depend, is itself thrown into question. The gap between the abundance of experiential possibilities beyond the system, of which the individual naturally becomes aware in the ordinary course of living, and the prefabricated behavioral possibilities that are made available within the system and that appear in modular forms of group behavior now becomes so large that it is not complexity, but rather enforced simplicity, which becomes the sustaining factor of existence for individuals.

This preference for familiarity over novelty is overtly referenced in the novel. Early on, a discrepancy is registered between the sense of an already lived real experience and the variety of experiential possibilities that are excluded from the Kakanian system. The bearer of this former state of mind is Ulrich's sixty-nine-year-old father, who, as the narrator tells us, is led in all ways by an inclination toward reality, that is, everything that is confirmed by the existing direction of his life. Ulrich, by contrast, does not take what has become familiar and real any more seriously than that which is merely thought of as possible and is not (yet) real. For mere thoughts, which can nevertheless contain "a possible truth," Musil deploys the phrase "sense of possibility" (*Möglichkeitssinn*). A person who puts trust in a sense of possibility is thus described as a person who lives "in a tangle (*Gespinst*) of fog, imagination, dreaminess, and subjunctives" and who not only contemplates what is not yet actualized but also treats the unactualized as a concrete "task" and an area of potential "discovery" (*Aufgabe und Erfindung*) in life (*GW* 1:16). Such an approach to reality turns away from the traditional categories of established notions of experience toward the abundance of possibilities that the social system tends to exclude. To that extent, "sense of possibility" means the speculative engagement with the complexity to which the system has already reacted allergically, so to speak. Since the Kakanian system focuses on reducing complexity, but systemically prevents the expansion of complexity, the sense of possibilities of the "possibility-person" (*Möglichkeitsmensch*) shifts into a realm beyond the experience of everyday life. A "sense of possibility" is thus something different than just a game (which is a structured experience found in everyday life). Musil makes it clear that possibilities that the narrative means to suggest as a concrete alternative to the available lived experience of conventional life must first take root in the imagination and, to use an expression of Albrecht Schöne (1961), be initially considered only "subjunctivally" available. From these beginnings, however, the search for possibilities becomes a search for elements of possible being in which the individual can place trust. This search demands a neutralization of everyday experience of reality, or at least the ability not to take reality more seriously than that which is not real.[8]

Musil's social critique also takes account of functional trust equivalents in social systems—semantic clusters associated with money, power, and truth that also feature in Luhmann's analysis. For Luhmann, trust in political power, in money, and in societally sanctioned "truth" are "typically decentralized forms of complexity reduction" that serve the needs of a general "simplification supported by the existing power arrangements." He then broaches the question of decision-making when the decisions in question "are not those that occur in the common run of things, but nevertheless come with the binding seal of legitimacy" (Luhmann 2000, 69–70). In this regard, Musil underscores how little binding authority these functional equivalents of systemic trust can engender for individuals. If the "glorious" attempt of the parallel campaign to give expression

[8] The critical aspect of this approach to reality, which since the work of Laermann (1970) and Schmidt (1975) has been discussed under the guiding concept of "characterlessness" (*Eigenschaftslosigkeit*), has generally stood out in the research. In contrast, the utopian orientation of the protagonist's search for possibility has been received with less unanimity, particularly in the socially critically oriented research of the 1970s and 1980s and since 2000 (in regard to the latter, see my discussion in chapter 5), because Ulrich's *Eigenschaftslosigkeit* has appeared to critics, at least in part, to be premised on aristocratic values that do not correspond to a strong form of social engagement or activism (see also Böhme's discussion, 1974). Nevertheless, the socially critical reading revealed precisely the factor of absent trust, which had largely escaped the attention of existentialist interpretations of the 1950s and 1960s. The central link between criticism of the untrue and non-authentic experience of reality that became the sustaining element of research in the late 1950s and 1960s, and the societal critique of the 1970s and 1980s is also not adequately dealt with in the aesthetic approaches that have enjoyed popularity in Musil research throughout the postwar period (to name just two examples, Hoffmeister 1965 and Wagner-Egelhaaf 1989). Perhaps that is ultimately because Musil neither aims to solve the social problems of the Kakanian system nor wishes to escape into theological fantasies or speculation. In contrast, he speaks of a stance that, in the modelling of alternatives to known reality, envisions a new reality, admittedly far off in the distance, but nevertheless (in his view) realizable. In other words, Musil is not a mystic, even if he was fully aware that a certain mystical quality attaches to the intimation of not-yet-realized possibilities for living.

to Kakania's inner "truth" is turned into a process of repetition in which only prefabricated material is to be rehearsed and actuated, then little actual truth will be the result. This is evident in Musil's portrait of the German industrialist Arnheim, a member of the parallel action in whom the "unification of soul and economy or of ideas and power" (*GW* 1:108) is embodied. But, as the reader soon learns, this "unification" is only the "front view" of the material order of things in Arnheim's imagination, what Musil labels "the large industrial form of the intellect" (*die Großindustrie des Geistes*): "It is of course life in the form it takes today that brings about the large industrial form of the intellect, just as, conversely, it is industry that rushes forward toward the intellect, politics, and control of the public conscience; both phenomena meet in the middle" (*GW* 2:429).[9] Arnheim gains the attention of the hostess Diotima precisely because of this "industrial" form of his thought and intellect. Ulrich, by contrast, whose thinking is of the opposite order—idiosyncratic and independent—comes across to her as immature. Even at the level of the individual, then, the industrial form of thought commands more respect than the private and the personal. Ulrich's fate on the organizing committee of the parallel action is thus to be passed over as aberrant presence, if not finally as entirely irrelevant.

Nevertheless, precisely at this point in the story in the latter stages of book 1 of the novel, Musil's critique of the Kakanian system comes into its own. Musil's counter position appears to contend that complexity reduction, though it might be strategized at the level of the social system, is not a restorative or salutary measure for individuals, who are always faced with the need to face outward toward complexity and live their lives and shape their goals as individuals. The fundamental stance of the protagonist, Ulrich, bears witness to this fact, to which the epithet "man without qualities" plainly attests. A "man without qualities" is certainly not someone who suffers from the inability to make a decision, or who only views the world and the task of acting in it negatively. A man without qualities opposes only those social processes and mechanisms that aim to build trust in ordered patterns that have nothing to do with individuals and that do not directly concern the feelings individuals develop toward their own life. What follows Musil's critique of the Kakanian system in book 1 of the novel, therefore, is a more idyllically premised second book in which the reflections of the two individuals brought into focus in this section—Ulrich again, but this time in partnership with his sister Agathe—aims, in an entirely different setting, to break

9 In the original: "Es ist natürlich das Leben in seiner heutigen Ausbildung, das zur Großindustrie des Geistes führt, so wie es umgekehrt die Industrie zum Geist, zur Politik, zur Beherrschung des öffentlichen Gewissens drängt; in der Mitte berühren sich beide Erscheinungen" (*MoE*, 429).

with the logic of the Kakanian system. In book 2 of the novel, in which an ideal of "inductive" communication is put forward as a counterweight to the "deductive" and deterministic reasoning of book 1, Musil develops a highly individual, non-systemic conception of trust that seeks to realize new possibilities and an entirely new sense of individual being in the modern age.

# 9: Musil's Correspondence Project: *The Man Without Qualities* and *The Blackbird*

In his fiction, his short prose works, and the major novel, *The Man Without Qualities*, Robert Musil experiments with "multivariate" correspondences.[1] As I endeavor to make evident in this chapter, these experiments may be understood as apparently random occurrences in the narrative material that build up to a different level of coherence when the factor of similarity or correspondence is applied. The significance of such similarity is only evident when it is appreciated how sharply the notion of multivariate correspondence, as Musil deploys it, breaks with the conventions and habits of traditional narration. This break principally serves the goal of enabling new modes of understanding and response to come into view that are commensurate with the increasing complexity of the world. Early versions of this correspondence project can be identified in writings such as the short story "The Flypaper" (1913) and the two novellas of *Unions* (1911), though these works will not be discussed here (the reader is instead referred to my discussions of these two works in chapters 6 and 7 respectively). In what follows, I will focus attention on two aspects of what I consider to be Musil's correspondence project: first, on passages from the major novel in which the essential form of this project is outlined; and, secondly, in the context of the later story *The Blackbird*, on a utopia of inductive experience developed from the idea of correspondence that is expressed in that work in a particularly striking way.

[1] My use of this term is taken from statistical analysis. For an introduction to multivariate statistics, see Anderson 1958. Franz Gustav Kollmann has undertaken a detailed assessment of Robert Musil's mathematical and statistical knowledge. While Kollmann acknowledges that Musil is certainly not to be understood as a mathematician, he did nevertheless acquire an unusual level of knowledge "for an author" in the fields of mathematics and theoretical physics. (Kollmann 2007, 71). This fact supports the conjecture that Musil had a good command of statistical analysis (ibid. 5).

## *The Man Without Qualities*

Musil's approach to narrative presupposes the primacy of form and function.[2] In this approach, a substantive element does not become visible, as it were, "inside out" from the viewpoint of a pre-stabilized subjectivity, as might be considered the norm in conventional narrative, where actors, at least notionally, are instances of, or become the conduit for, sovereign narrative experience. Rather, attention is directed away from these putatively sovereign actors or "protagonists" as the gaze of the narrative begins to settle on external events and the contextual features with which these actors contingently align. This approach to structuring the narrative is already discernible in the first sentences of *The Man Without Qualities.* Musil does not speak in these first sentences uncomplicatedly, as would be typical of a conventional novel, of "a beautiful August day." Instead, his narrator refers to "a barometric low that migrated eastward toward a high positioned over Russia." This low "did not yet betray the tendency to evade this high by moving north." Details about air temperature, the "monthly temperature fluctuation," the "tension" of the "humidity in the air" and other weather-specific factors are adduced in order to "convey the facts properly" (*GW* 1:9). Only after the provision of such a plethora of meteorological data, which furnishes the narrative perspective with sufficient factual detail about this particular day on which certain other events occur, perhaps noteworthy, perhaps not, does the narrator see himself in a position to reach an equivalence with conventional narration, that is, to use the "old fashioned" formula: "It was a fine day in August in the year 1913" (ibid.).

What content is in the narrative therefore arises, first, from an experimentalism practiced in, and discharged through, the technical—one of the new languages of the world in the sense established in the introduction to this volume. The basic outlook of narration is thereby entirely reimagined. Instead of a conventional narrative mode where expectations are pegged to a coherent intellectuality in the "linear" space of the subject as protagonist, an experimental field opens up according to which functional correspondences become recognizable across ostensibly unrelated semantic assemblages—for example, the technical language of meteorology becomes functionally aligned with ordinary perception and to some extent takes precedence over it. The result is that a basis for a procedurally "subjectless,"[3] non-linear correspondence is established. The value of

[2] Musil's deployment of the functionalist approach is treated in detail in chapters 1 and 2 of this volume.

[3] The similarity of this subjectless process to systems theory is obvious (see also note 2 above). Systems theory, which presumes the primacy of the fact of society before that of the individual or of the subject, exhausts itself in an *extensive* systemic analysis and has in my view no basis to thematize the *intensive* conditions of

linear narrative founded on the primacy of subjective consciousness is no longer clear. This is primarily because what is real is no longer shown to be subject-enabled, that is, no longer takes on forms that follow a coherent view of objects[4] anchored in subjectivity as a reliable source of information about the world. In place of this "univariate" object view, whose logic stems from and presupposes the controlling instance of subjectivity, a multivariate procedure emerges in variably contingent emergent fields whose functional entities process detail by moving, as it were, "outside in," that is, from an external position outside human subjectivity toward points in common with them, though without foreknowledge of how the constitution of such a subject is actually to be imagined. What an entity is—according to this perspective—depends on the input settings of correspondences such as they might come into view. As such, when Musil employs the language of meteorology in the first paragraph of the novel, it is not only for the purpose of cueing an ironic narrative intention. This language simultaneously stands for a method of seeing that understands the environment of the narrative as a statistically variable field, the structures and forms of which are subject to a potentially endless variability, and thus appear "multivariate." We have already seen in chapter 6 in relation to our analysis of the short prose text "The Flypaper" how this subjectless view overlaps with the perspective of posthumanism. From this angle, Musil's "multivariate" outlook anticipates the arrival of the posthumanist paradigm in today's cultural discussions.[5]

The implications arising from this kind of narrative are further in evidence in relation to a traffic accident that the narration soon remarks upon. An unnamed pedestrian—a denizen of the new age of modernity that serves as the novel's setting—witnesses an incident in which a cyclist is apparently knocked over by a truck. The first words of this pedestrian continue the "outside-in" perspectival shift already highlighted. The pedestrian observes: "These heavy lorries have a braking distance which is too long" (*GW* 1:11). Shortly thereafter he reports statistics relating to casualties from traffic accidents in the United States. The female companion of this pedestrian feels "relieved" at this statistical portrayal of the

---

the ego-experience. It is precisely the latter conditions that are (also) the subject of the correspondence analysis presented here. On the problem of the absent subject in systems theory, see Pott 2013,193–207. Note also my detailed application of systems theory to *The Man Without Qualities* in chapter 8.

[4] Such an approach would be "univariate" when it understands an object as an invariable singular phenomenon. In contrast, an approach is "multivariate" when its objects show themselves in constantly new object-forms, and as such possess ambiguous interpretive values.

[5] A well-informed recent study of posthumanism in the context of German literature and thought can be found in Landgraf et al. (2019). For a discussion of the relation of science to posthumanist debates, see Barad (2007).

event she has just witnessed. She thanks her companion "with a thoughtful look" because this technical explanation of the accident saves her from having to deal with her feelings about what has just occurred.

Musil's experiment in the major novel—and in many other works—consists of drawing out the implications from this radical perspectival shift: on the one hand, the impact of external events and processes is strengthened, while, on the other, the authority of perspectives associated with human subjectivity is weakened. The point of this inversion of perspective appears to be that the viewpoint of subjectivity per se has become, or must now be considered, indeterminate. This aspect becomes clear in relation to the two human figures on whom attention focuses in this first chapter. They *might be* the two characters who appear later in the novel, that is, Arnheim und Ermelinda Tuzzi, or they might not be. The narration refuses to say. About this situation of indeterminacy in the urban spaces of modernity only this, apparently, can be said with any confidence: "Lively people (*lebhafte Menschen*) sense such puzzles quite often when they are out on the streets" (*GW* 1:10). What does the narrator mean by this? Are these "lively people" a special class of people who always have their wits about them and remark upon what they see when they are wandering the streets? Or are they people like the narrator, people whose attention serves the purpose of writing stories? These questions remain unanswered. What we do learn from the narrator about these two figures, however, is nevertheless something surprising—the fact that they had "their initials embroidered on their garments, and in this way, that is, not turned outward but inscribed on the fine undergarments of their consciousness, they knew who they were" (ibid.).[6] One can take this observation simply to mean that these two figures were quite sure in their own skin of their own identity. For a correspondence analysis, however, which focuses attention on the loose couplings that are intrinsic to occurrences and the accidental alignments that come into view, the observation can be taken to imply that identity is in fact conferred "outside in," by some aspect of outer reality that, whether through accident or design, is taken over into the inside, where it melds with subjective experience and establishes subjective identity.

If we follow the prompts that are given in this opening chapter, we are led, I believe, to a problem of subjectivity.[7] This problem has two aspects. While some citizens of Kakania, as the just quoted passage indicates, obviously think they know only too well who they are, other citizens—the protagonist of the novel is one of them—patently lack this

[6] Musil's original: "die Anfangsbuchstaben ihrer Namen bedeutsam auf ihre Wäsche gestickt, und ebenso, das heißt nicht nach außen gekehrt, wohl aber in der feinen Unterwäsche ihres Bewußtseins, wußten sie, wer sie seien."

[7] Stefan Jonsson provides a clear description of this problem (2000). For a psychoanalytical interpretation of the problem of subjectivity, see also Jappe 2011.

certainty. The formal preconditions for Kakanian living therefore appear not to provide a sufficient basis for apprehending the world from assumptions rooted in subjects and the traditionally subjective. For this reason, the intellectual atmosphere of the novel presents as a transitional state in which old habits of mind and received ways of being and understanding identity no longer provide adequate orientation, while new ways of orientation and new expressions of future being are not yet available. This transitional state is further underscored by the problem of complexity management discussed in the previous chapter: when no state of mind or being seems adequate to the task of complexity management or of complying to the agreed terms in which it has been brokered, then process (as the problem of dealing with complexity) inevitably trumps "substance" or "content" (the object view of conventional narration arising from a coherent enabling subjectivity). The title of the novel already suggests a consequence of this in-between state. The character named in the title, at least to all appearances, is a featureless man who, whether by accident or design, eschews the stabilizing effects of established human qualities or attributes. The featureless condition of the absence of qualities seems necessarily paired with the waning of a strong sense of reality—the condition in which humans no longer have "sufficient reason" for fashioning their lives according to reference to a particular set of "qualities" or attributes by which subjective identity is made.[8] The question of what social, cultural, and intellectual circumstances must emerge in order to end this state of transition and allow for the arrival of the new is one of the most pressing concerns of the novel.

The tension that the narration develops, for this reason, is the contrast between *Wirklichkeitssinn* and *Möglichkeitssinn* discussed in previous chapters.[9] From the vantage point of correspondence analysis, these terms function as glosses for two contrasting strategies of world-comprehension: respectively, the ***univariate***, focused around a single view of the world with pre-patterned components and simplified modular outlooks and a

---

[8] Cf. *GW* 1:131: "It could even be maintained that [middle-aged people] had been deceived, for one can nowhere find a sufficient reason for the fact that everything happened as it did; it could equally have happened differently. . ."

[9] Musil's concept of possibility, and the aesthetic goals that he thereby pursues, have been illuminated by Thomas Sebastian in a notable study of Musil. In Sebastian's view, the concept of possibility serves the formation of a "hypothetical narrative" whose primary purpose is to imaginatively cut a path through to other possibilities of being. As Sebastian says: "In such an explicit counter-factual narrative, passing from one possible event to the next becomes a question of probable causality. It depends, so it seems, both on the observer's knowledge of the hypothesized world and on the plausibility of the conclusions drawn from that knowledge. The story's suggested course should not contradict the facts implied in the original hypothesis" (Sebastian 2005, 89).

concentrated, hierarchical flow of executive power; and the *multivariate*, focused on a plural, diversely layered, open view with dynamic components and assemblages and emergent, "complexified" outlooks, where power is disbursed by way of irregular multidirectional energy flows throughout the social system. Both strategies are associated with particular characters in the novel: Clarisse, Walter, and Ulrich—the "quality-less man" of the novel's title—are the main carriers of multivariate reference in the novel. By contrast, Arnheim, Diotima, and Ulrich's father are the figures whose manner of relating to the world can be considered univariate and for whom the exercise and flow of power is "top-down," unidirectional, and maximally concentrated in the hands of a few. Book 1 of the novel is built on the intellectual contrast between these two forms of comportment. This dualism is ended, albeit only temporarily and provisionally, when the father—whose limited role in the text is cued by letters and memories, not by his specific appearance as a character in the novel—withdraws completely from the plot upon his death. By these means, the attention of the narrative is drawn progressively to the question of social and intellectual potential that might be attached to multivariate comportment. Is the multivariate approach to world-comprehension and to the problem of acting in the world able to counteract the centralizing tendencies in modernity of which Kakania provides a model? Is a sense of coherence built from emergent, rather than already established properties and qualities, a plausible source of orientation in the world? If, as the Other Condition in the second book of the novel appears to suggest, such a source of eccentric movement of both sense and purpose in the world is indeed possible as an alternative way of living in the world, how does it work in practice?[10]

These are the type of questions that arise in book 2 of the novel. The "container" for these questions is the Other Condition, and the vehicle or agency that drives the Other Condition and brings it to light as a practical idea is an unusual love between the siblings Ulrich and Agathe.[11] The first moment of recognition of this love, in keeping with the strategy of multivariate correspondence followed by the author, occurs under the presumption of contingency. At their first encounter, the two siblings—who have not seen each other in a very long time and have accordingly become estranged—are similarly attired. As the narrator reports, Ulrich found himself "through a mysterious ordination of fate faced with a large, blond wearer of pyjamas that were covered in delicate, gray, and

[10] The motif of transgression as a life strategy and intellectual principle in the sibling idyll leads, as in Hajduk's interpretation of *Der Mann ohne Eigenschaften*, to a principle of aesthetic border-crossing as an "erhabene Grenzerfahrung des Selbst." See Hajduk 2000, esp. 355–89.

[11] An overview of the ethical orientation of the sibling idyll can be found in Mehigan 2001, 128–36.

reddish-brown stripes and cubes" (*GW* 2:676).[12] As the narrative makes clear, certain random variables in external events—that the siblings are wearing the same type of pyjamas—underscore the fact of what turns out to be an unusual correspondence in many areas. Nevertheless, it is also the case, as noted by Agathe, the female element in this unexpected encounter, that something more than accident is in play here: "You can say it was chance; but what is chance? I think it is fate or providence or however you might call it. Has it never struck you that it can be put down to pure chance that you were born as who you are? That we are brother and sister means that you can double that!" (*GW* 3:904).[13] Agathe addresses here what might be understood as a mix of contingent emergence and mythical convergence. Musil's experiment provides for conditions that presuppose the endless variability of the surrounding world, but which fundamentally assume the possibility of a redisposition of this variability toward an intensification of the lived sense of the world for individual actors. In line with this assumption, it is possible to see the sibling idyll as articulating an insistent demand for "a *doppelgänger* of the opposite sex," and a desire for "the love of a being who is identical to us but who is also other than we are, a magic creature that is us, but that actually remains such a creature and that above all, more than we can even imagine, breathes [the spirit of] autonomy and independence before we do ourselves" (*GW* 3:905).[14]

## *The Blackbird*

The story *The Blackbird* (*Die Amsel*) was published in 1928 in the journal *Die Neue Rundschau* and was also included, albeit with several amendments,[15] in a collection of short prose Musil published in 1936 under the ironic title *Posthumous Papers of a Living Author*. In this story, Musil's deployment of correspondence strategies takes on even more pointed significance than that already observed in relation to *The Man*

---

[12] Musil's original reads: "durch geheime Anordnung des Zufalls einem großen, blonden, in zarte graue und rostbraune Streifen und Würfel gehüllten Pierrot gegenüber" (*GW* 2:676).

[13] In the original: '"Du kannst sagen, dass es Zufall gewesen sei; aber was ist Zufall? Ich glaube, gerade er ist Schicksal or Schickung oder wie du das nennen magst. Ist es dir nie zufällig vorgekommen, dass du gerade als du geboren worden bist? Doppelt so viel ist es, dass wir Geschwister sind!"' (*GW* 3:904).

[14] The passage in Musil's original reads as follows: "die Liebe eines Wesens, das uns völlig gleichen, aber doch ein anderes als wir sein soll, eine Zaubergestalt, die wir sind, die aber doch eben auch eine Zaubergestalt bleibt und vor allem, was wir uns bloss ausdenken, den Atem der Selbständigkeit und Unabhängigkeit voraushat" (*GW* 3:905).

[15] On the development of this text see Arntzen 1980, 163–64.

*Without Qualities.* If, in the latter, Musil was interested in assaying oppositions in order to critique traditional modes of thought and perception, in *The Blackbird* Musil aims at a simultaneous doubling and fracturing of perspective that takes the correspondence achieved to a heightened level of significance. By these means, the narrative perspective breaks into two paired, but also sharply divergent, paths of meaning-making that recur to the distinction established early on in *The Man Without Qualities* between *Wirklichkeitssinn*, a type of meaning-making that only seeks to ratify the existing state of world-reality, and *Möglichkeitssinn*, a making of meaning that seeks to depart from the existing state of world-reality in its attempt to imagine entirely new possibilities for living. This distinction is aligned in *The Blackbird* with the perspectives of "A1" and "A2" respectively.[16] Whereas the satiric intent of book 1 of the major novel was such that the separation between these two ways of seeing the world was condensed to merge with particular characters, in *The Blackbird* these considerations are completely abandoned in favor of an exploration of the creative possibilities associated with, and arising from, the stance of *Möglichkeitssinn*, the sense of possibility, alone. For this reason, greater weight in *The Blackbird* is accorded to the viewpoint of A2. Given that the ambition of book 2 of *The Man Without Qualities*, which Musil was not able to finish during his lifetime, lies precisely with the promise intrinsic to this stance to offer a different way of approaching the real and imagining the possibilities of life that modernity affords, it is possible to view *The Blackbird* as an encapsulated version of the virtues of the Other Condition that is put forward in book 2 of the major novel.

The contrast Musil's correspondence strategy aims at is introduced at the very beginning of the story in the following way: "Friendships struck in youth seem more and more strange the older one gets" (7:548). Why that might be the case is soon linked to an even deeper problem, namely how one maintains continuity between the "various personnages one successively addresses with the term 'I'" (ibid.). This second problem is important not least because of the intellectual modifications that take place as the development of physical life occurs. These changes are expressed—in accordance with the "outside-in" view of creaturely life already set out in the previous section of this chapter—as the process by which an externality impacts upon a notional subject internally and brings about a shock in the lived ego-experience of that subject. The question posed in *The Blackbird* is therefore the extent to which an inner sense of the "I" can be sustained in the face of external events that necessarily

[16] I refer to these figures with numerals rather than, as Musil does, with numerals written out as words (i.e., "Aeins," "Azwei"). The use of numerals helps us see the modular intention in these names and the fact that both can be viewed as variants of the same basic identity ("A").

occur contingently and, in the first instance, are necessarily processed as jolts and shocks in the lived experience of the self. How, then, does random variability in its most concentrated effects modify the substratum of the I or self without dividing or rupturing it and terminally comprising that substratum?

This problem of continuity at the meeting point of internal experience and external contingency—the problem at the core of this story—is fundamentally linked to a general problem of narration. Taken literally, this is the problem of the extent to which a meaningful overall picture of events can be produced out of the prevailing complexity of the world through the inner workings of narrative. This problem comes to the fore at the end of the story in the exchange between A2, the first-person narrator, and A1, his skeptical questioner in the narrative. A1 wants to know whether "all this"—the three remembered stories from A2's life—"are linked," or, as the German original says literally, "has a common meaning (*gemeinsamen Sinn*)" (*GW* 7:562). A2 takes this question to be a question about whether narrative admits of closure, that is, as a hermeneutic question about how meaning arises in narrative. Can the sense of narrative be provided for "internally" within the narration, by those who are the narrators? In response to this question, A2 is completely unable to say where this "common meaning" lies, but his response is nevertheless instructive: "If I knew the meaning [of what I have narrated], I wouldn't need to relate [these stories] to you" (ibid.). We will return to this quotation at the end of this chapter. For now it can be established—at least this is the proposition that is explicitly entertained in *this* narrative—that states of the ego and the narrated event-moments of narratives might indeed have something "in common." Musil's correspondence strategy provides a fruitful lead in this regard. This is a strategy where the task devolves to narrative of configuring alignment between the "event horizon" of perspectives internal to a subject and the notionally random events on the impersonal plane of outer life. What Musil's correspondence strategy appears to lay out, then, is a problem of induction by which the test of the coherence of inner states of being—the "identity" of a particular subject or that subject's claim to coherence "on the inside"—has an analog in the capacity of narrative to generate its own conditions of "narratability."

In A2's three remembered stories unexpected events are encountered. As already established, a narrative strategy of "inside to outside"—the vector of traditional narration—is unavailable as a means by which to organize or process the encounter with these events. Instead, an "unmade" or, perhaps better, unfinished interiority is put before a random field of contingent events in outer reality without any connectedness with these events necessarily arising. Of course, the title of Musil's work prepares us for some sort of link between these events; for this reason it can be proposed that the events that impinge on A2's ego-experience provide instances of an encounter with a thing or being that both flies and sings. This flying and singing thing is perceptible only to A2 and, accordingly, is private to his internal viewpoint. In the first story it is a nightingale which turns out to be a blackbird, in the second it is a bomb shard whose musicality as it descends to earth is only audible to the narrator, despite the presence of other soldiers, and, finally, in the third story, it is again a blackbird, but the song of this bird reveals itself to be the voice of A2's recently deceased and interred mother.

In the first story the narrator wakes in the middle of the night and hears the song of a nightingale. He quickly reckons that nightingales are somewhat rare in inner Berlin, and as such it must be a "bird from heaven"—one of those creatures encountered in the magical world of childhood. Since, at that time of night, his wakefulness was "different from that of day" and, as such, sensitized him precisely to the type of objects and events encountered in such a magical world, he found himself "prepared in the most natural way to believe in the supernatural." At precisely this moment, moreover, he resolved upon a step of immense consequence for his personal life: "I will follow the nightingale" (*GW* 7:552). In the second story, which is set on the front in South Tirol during the First World War, an "enemy dispatch" suddenly disturbed "our peaceful position." Initially, the narrator heard only a "quiet ringing" that apparently his comrades could not hear. It was the narrator's first experience of a bomb shard, and "at the next moment I had a strange certainty not founded on any sense of probability: it will hit me!" (*GW* 7:556) In the third story-within-a-story, which relates circumstances pertaining to the death of the narrator's mother and about the days A2 spent disbursing his parents' effects (his father died soon after his mother), it is once again a bird to which the narrator's attention is drawn. Buried during the day in children's books, which brought his forgotten childhood alive again for him, A2 was woken after midnight by "a wondrous, splendid singing." In his sleep he perceived this singing to be the song of a nightingale, but upon waking he immediately knew it to be a blackbird. The blackbird, as A2 then recounts, speaks to him: "'I am your mother—she said.'" Just as he did long ago in childhood, the narrator fetched a wooden cage for the bird, fed it, and allowed it to fly freely in and out of an open window. Then the narrator came to this insight: "'I was never a better person in life than when I had the blackbird; but I probably can't tell you what a good person actually is'" (*GW* 7:562).

Uncertainty about "what a good person is" prepares the reader for the same sort of ethical problem Musil discusses in book 2 of *The Man Without Qualities* (as suggested earlier, *The Blackbird* foreshadows the direction the great novel embarks on but could not bring to conclusion). Musil's correspondence strategy, which discusses the nature of disparate life experience, casts some light on the nature of this ethical problem. The blackbird—or more properly, the "heavenly bird" connecting each of the three stories-within-a-story—is not comprehended as an object promoting transcendence but rather as a chance materiality from which the psychological achievement of continuity between ego-conditions is gained, a continuity which consists in the identification of moments of ethical significance in the subject's life that become recognizable and linked in the substratum of the ego. The heavenly bird in each of the three stories thus hints at formations in the ego that accrue inner ethical sense. Noteworthy is the fact that the first person narrator has no direct access to such inner sense, nor can it be rendered intelligible by way of explanation or description to an interlocutor (represented by the perspective of A1 in the story). Instead, access to this ethical sense must be sought as a "third possibility"[17] of the narration in a hermeneutical space[18] at the level of textual reception. In the symbolic language the narrative develops in the conjuring of this space, the ethical is decipherable not through the primary sense of sight in the plain light of day, but via the secondary language of aurality under conditions where light is absent or sight on its own is insufficient—the birdsong that pierces sleep, for example, or the rapidly moving bomb shard that is too small to see but can be heard. As A2 puts it, it is "as if you could hear whispering or merely a rustling, without being able to distinguish between them" (*GW* 7:, 562).

The ethical in *The Blackbird*, accordingly, inhabits the domain of the in-between, where one might hear a rustling and not take it to be a whispering, a circumstance that, for example, would map onto A1's standpoint, or equally, the other way round, a rustling that is also a whispering, an invitation to take the sound one hears to be ethically meaningful—A2's position. In terms of the first story, in which Musil's correspondence strategy begins to scale up to ethical significance, this amounts to a decision "to follow the nightingale" in full knowledge of the slippage that makes it improbable that an actual encounter with a nightingale has taken place. In the middle of a restless night in his apartment in Berlin, the narrator uses this other type of hearing as an invitation to take a step away from the settled circumstances of established bourgeois life. He resolves to leave his wife. A rational calculus applied at this point would doubtless have counseled against such a step; that some measure of personal injury and disruption to the life of the man's female companion, if not also to his own, would flow from such a step can hardly be doubted. It therefore belongs to the view attaching to these moments of unusual or special perception that ethics is to be defined as that which concerns the direction of individual life and can only be accounted for "ethically" in reference to that life.

[17] See *GW* 2:582, where the exclusion of a "third possibility" is discussed in connection with the problem of securing meaning for the life one leads.

[18] On the closeness of the correspondence analysis represented here and the discussions of space in literature, which likewise involve the theoretical designs of similarity-thinking, see Bhatti et al., 245n7.

In the second interpolated story within the story the narrator also throws over bourgeois normality. It is completely unexpected that A2 can perceive the little arrow of a shard that hurtles his way and consequently seriously maims him "not as a terrible foreboding, but as an unexpected moment of happiness" (*GW* 7:556). Why this might be the case arises from the narrator's fundamental openness in relation to the objects of the external world—a stance which provides for the possibility of correspondence between random occurrences on the outside and the private form of ethical life on the inside:

> And this sound was directed toward me: I was connected with this sound and had not the slightest doubt that something of extraordinary significance was about to happen to me. Not a single thought in me was like anything resembling the last moments of life; what I experienced was directed to the future, and I must simply say I was certain of feeling the imminent nearness of God near my own body. (*GW* 7:556)

Of course such intimations of significance in external events could be construed as completely irrational. It is here that the caviling voice of A1 comes into its own as a skeptical presence that opposes each moment of extension into the ethical associated with the stories of A2. By these means, the denial of significance is expressly provided for. A1 therefore asks, almost cynically, whether the blackbird of the third story "often spoke to him" (*GW* 7:562). His words imply that A2 has turned into a mystic. The sense of a cynical counter-view gains ground when one keeps in mind details about the friendship between the two mentioned at the beginning. As we are told, both friends during their time at university had defended a "materialistic worldview that looked upon the human being as a physiological or economic machine lacking a soul and a god, which that human being perhaps really is" (*GW* 7:549). Of the two companions since then, only A1 had kept up such a view. This alone would support the element of scornful regard that has crept into their relations since that time. Meeting again after a long period, each thus appears to question the stance of the other and seek validation of their own position. Their encounter implicitly raises the question of whether A1's austere, skeptical way of life is ultimately to be preferred to the more intuitive, less rule-bound and regulated, and ultimately more individuated way of living promoted by A2.

Though A1 is the more rational of the two friends and his skepticism appears a valid response to the stories he has heard, it nevertheless appears clear that it is A2's standpoint that is commended in *The Blackbird*. The author's intention with the story, as we have seen, is to draw attention to, and delineate the terms of, a way of response to the world premised on the sense of possibilities—a precursor to the state of the Other Condition in the major novel that was already in an advanced state of preparation in 1928 when *The Blackbird* was published. Here the attempt is made to understand the task of living—formulated in the major novel as the question of *das rechte Leben*, the "right way to live" (*GW* 1:255)—as a question of the shape of inner life where living and writing, ethics and aesthetics are imagined in partnership. The clearest expression of such a merging of ethics and aesthetics, as the interpolated stories of *The Blackbird* reveal, is to be found in a childlike form of response to the world. As the narrator explains, childhood "at both ends is not completely secured." A child has neither a cool and calculating mind, nor fully-grown legs, though exactly this proves to be an advantage, because "instead of the pincers of later life one still has soft flannel hands . . . and can sit in front of a book as if one were sailing through space . . ." (*GW* 7:561). Similarly, in *The Man Without Qualities*, it is this not yet ossified nature of the child's mode of understanding to which the novel's protagonist would like to return:

> Nothing is quite there any longer as it was in childhood. But as soon as you have succeeded in becoming a "personality," everything you touch is, comparatively speaking, frozen, right up to your core, and what remains, covered in a thoroughly external being, is a ghostly wisp of foggy self-assuredness and an opaque love of self. Isn't there something that's not quite right here? One has the feeling that at least something should be reversed! It's not enough to merely to say that a child experiences things differently from a man! (*GW* 3:902)[19]

Musil's picture of the virtues of the child and what might be called an ethic of childhood appear as the vanishing point of his strategy of correspondence. This strategy is ethically informed and attempts to realize a widening of perspective beyond the settled "univariate" view of objects with which it is contrasted. Musil's strategy does not allow an object to be an object in any pre-established way. Musil instead casts the external world as a constantly shifting, "multivariate" field of contingency in which coherence is yet to appear or be ascribed. Coherence strategies can be ready made and established, as would be consistent with the stance of *Wirklichkeitssinn*, or open and unmade, as with the stance of *Möglichkeitssinn*. This latter stance, which Musil links to the unfinished thinking of a child, only retrospectively allocates meaning to the events it experiences. For example, the various "blackbirds" that appear in *The Blackbird* first forfeit their

[19] The original German is as follows: "Nichts ist mehr ganz so da, wie es in der Kindheit einmal gewesen ist. Sondern es ist alles, was du berührst, bis an dein Innerstes verhältnismäßig erstarrt, sobald du es erreicht hast eine 'Persönlichkeit' zu sein, und übriggeblieben ist, umhüllt von einem durch und durch äußerlichen Sein, ein gespenstiger Nebelfaden der Selbstgewißheit und trüber Selbstliebe. Was ist da nicht in Ordnung? Man hat das Gefühl, irgend etwas wäre noch rückgängig zu machen! Man kann doch nicht behaupten, daß ein Kind ganz anders erlebe als ein Mann!" (*GW* 3:902).

ordinary meaning. The correspondence of the blackbird and the nightingale, and thus also of a meaningful emotional factor arising within a person, is not gleaned from a scientific approach. The scientific viewpoint would be accessed through the ornithological insight that blackbirds typically imitate other birds. As much as such a viewpoint might in other circumstances be advantageous, in terms of the correspondence strategy put forward in *The Blackbird* it is of little value since it is unable to provide an ethical way to live. What instead is commended is an approach to living that is rooted in the type of deeply personal decisions a human being is called upon to make when that human being lives authentically as an individual. That such an individual appears in a position to advance ethically is due to their openness to the external world—an openness consisting of the willingness, say, to interpret the sound of a blackbird as the song of a nightingale, despite the fact that it would be rationally indefensible to do so. According to Musil's radically individual ethics, random manifestations may count more than rational assessments, despite the forcefulness of the general claim of such assessments to attention in modern life.

Nevertheless, in Musil's correspondence strategy, at least as it appears to have been meant in both *The Man Without Qualities* and *The Blackbird*, the conduct of ethical life insists on a turn away from scientific ways of knowing and a turn toward the lived experience of individual subjects. As I have attempted to demonstrate in my discussion of the story *The Blackbird*, it is to narrative that the task of extending, intensifying, and deepening the ethical direction of individual life devolves. Just what such a narrative ethics that seeks a level of maximal openness to the events and occurrences of outer reality achieves on the inside, however, proves difficult to define. The task of making clear what such an ethics achieves and how finally it is to be thought of defaults in the end to what is described as the "third possibility" of the narration. Let us extend from Musil's narrative and call this possibility "A3." In this position of possibility we are confronted with a task of reception—a hermeneutic task of clarifying and determining the circumstances of ethical life as they might present to a particular individual. An ethics of these proportions is of course never ending. To A1's question about whether "all this" has one common, final meaning, A2, for this reason, can only put off the answer: "Good God—A2 countered—it all just happened like I said; and if I knew what it meant I wouldn't need to relate it to you. But it's as if you could hear whispering or just a rustling, without being able to tell the difference" (*GW* 7:562).[20]

[20] In the original: "Du lieber Himmel,—widersprach Azwei—es hat sich eben alles so ereignet; und wenn ich den Sinn wüßte, so brauchte ich dir nicht erst zu erzählen. Aber es ist, wie wenn du flüstern horst oder bloß rauschen, ohne das unterscheiden zu können!" (*GW* 7:562).

# Conclusion: "A General Secretariat of Precision and Soul": Ethics, Knowledge, and Literature after the Fourth Revolution

FEW WOULD DOUBT that the terms of knowledge have undergone a fundamental change in recent times. In the view of many scholars who have tracked this shift,[1] we are currently journeying through a revolution that is changing the way we arrive at knowledge. This is not a revolution about what humans know, as with all three previous revolutions. In this latest revolution, a so-called "fourth revolution" in knowledge, knowing has moved in significant measure away from the agency of human beings on account of the technical contrivances that help us know things. Our knowing, on this view, is fundamentally altered by the machines we build—above all the digital computer. These machines become a factor in the generation of knowledge not least because they exist outside us and outstrip the innate capacity of human beings to arrive at knowledge on their own. As a result, the ideal of "unassisted human reason," once cherished by the "philosophes" of the eighteenth century, no longer pays off.[2] "Big data" is the headline of a front-page story whose backstory tells of how the vast memory and processing capacities of modern computational devices help us know things we would never otherwise come to know if left to our own devices—these devices being, say, the Kantian time-space parameters governing human perception and cognition and the schematism that helps us know things from the tiny molecules of proto-knowledge that our sensual hardware and our conceptual software make available to us.

The new aspect of this knowledge revolution is not anything that we now know differently—for example, according to a genealogy promoted by Floridi (2014), that the earth revolves around the sun and not the other way round (the first, Copernican, revolution), or that humans descended from apes as a result of the workings of the same processes of evolution that govern the emergence of every other life-form we encounter on earth

---

[1] Some of the most prominent of these theorists are Floridi (2014), Brynjolfsson and McAfee (2016), and Schwab (2107).

[2] Kramnick, in his lucid introduction to *The Portable Enlightenment Reader*, draws attention to the interest of the Enlightenment in this type of reason (1995, xi).

(the second, Darwinian, revolution), or even that our wakeful consciousness is swamped by unconscious stirrings, attachments, and projections that further destabilize our capacity to have reliable "unassisted" knowledge of things (the third, Freudian, revolution). Rather, with the fourth revolution, it is not the case that we know big new things so much as that we know them by different means, and, if we concede the importance of this changed mode of our knowing, we can no longer place ourselves in all ways at the center of the knowledge enterprise, for human knowing is only one way to know things, and it has notable species limitations. Whether or not this fourth decentering of the human being, this further "descent of man," might turn out in the fullness of time to render us at a cultural disadvantage in relation to the machines we build in the pursuit of knowledge is a question that the literature of artificial intelligence and the "fourth revolution" now puts before us.

## Musil's Thinking in Light of the Fourth Revolution

In what follows in this concluding chapter, I aim to discuss this posited fourth revolution in knowledge in the context of ideas Musil develops in all his works but brings to a point of culmination in the novel *The Man Without Qualities*. As I contend, this novel anticipated the arrival of the fourth revolution in knowledge by several decades. Yet even more noteworthy than its prescience in this respect is its attempt to set out a comprehensive response to the new terms of knowledge—a response that does not maintain that the new knowledge will only be bad for us, nor that human beings are on a path that will inevitably lead to their ruin. On the contrary, Musil holds the path into the future to be positive so long as the virtues of modern knowledge can be understood and the downsides attaching to it avoided. Musil's particular confidence about the future concerns the prospects of intensifying the experience of individual life in new and more open ways than have been possible hitherto. The novel, from this angle, sets out what amounts to a new understanding of the opportunities of living in modern times and, underlying it, a new way of thinking. The parameters for this type of thinking are put before the reader in chapter 116 of the major novel, entitled "The Two Trees of Life and a Proposal to Establish a General Secretariat of Precision and Soul." This chapter represents the climax of the long first book of the novel. Embedded in the chapter's title and the concerns it introduces are both a problem and a solution. The title identifies the problem of a separation in thought running throughout Musil's works that we have discussed at length in this volume: the separation between the "two cultures" of science and art. The novel's protagonist, whose direct experience of this separation is central to the novel's purpose, puts forward a creative solution with a suggestion about how these two aspects of life might be brought into more fruitful dialog with each other. This is a suggestion that aims to be of benefit to society and to address the ailments from which this society appears to be suffering. The linking idea between book 1 and book 2 of the novel is the assumption that the new type of thought that is needed will have to go beyond the scientific and the technical. This is the new idea that the protagonist proposes to the organizing committee of the "parallel action" in chapter 116.

To appreciate the full dimensions of the problem Musil identifies, we must enter into the soul and psyche of the main character, a mathematician. We are told about this mathematician, who is an intellectual in every sense of the word, that he has taken what amounts to a "gap year" in order to come up with answers to certain weighty issues that exercise him. We are further told that he has even harbored thoughts of suicide in view of these issues. The reader cannot quite determine whether we should condemn or lament this character for these thoughts, privileged as he appears to be in every other way, and this sense of interpretative indeterminacy about a character who appears to blow hot and cold by turns is balanced against an equal indeterminacy about the true meaning of the novel's title. What is a "man without qualities"? Is that a good thing or a bad thing? Over the hundreds of pages that build toward chapter 116 in the novel we are brought to the realization that this man without qualities is not any conventional human subject, but the bearer of an entire cultural development. Thus, while we see a world largely from the perspective of this individual, we also see that his travails and afflictions are cast against a much broader canvas. In chapter 116 the narrator "doubles down" on these travails and afflictions, seeing the problem of the "two trees of life" as being at their core.

The problem is introduced as the main character's awareness of two conflicting aspects in his psychological makeup. These two aspects—deep drives in his personality that serve to explain his tendency to blow hot and cold in his life—are the drive to violence on the one hand, to love on the other. Violence, though we might associate it with something vital, even choleric, is in fact the cold element in his temperament. A quote from the novel illustrates this point: "It [the word violence] meant an outflow from every kind of skeptical, factual, conscious behavior; a certain hard, cold aggressiveness had even entered into his choice of career, so that an undercurrent of cruelty might have led to his becoming a mathematician" (*GW* 2:591). We have less difficulty in recognizing the drive to love in Ulrich's character as a hot drive, but, as the novel soon reveals, there is no Spinozist joyfulness attaching to it, since the love affairs that ensue provide at best only passing pleasure, and, more often than not, a good deal less than this, and thus only increase his dissatisfactions. And it is these dissatisfactions, above all, that have led to the awareness that he, like the society around him, is weighed down by torment and a proliferating cultural malaise.

Musil was uniquely placed to appreciate the importance of the subtle shifts in the intellectual landscape of his age that lay behind the emergence of this cultural malaise. As an engineer from an academic family (his father headed a department of a technical university), Musil witnessed at first hand how an increasingly sophisticated science of probability was changing mathematics and how the advent of early quantum mechanics was changing the physical sciences. There are rich notes about these matters in his diaries. As a psychologist, he followed closely early work in Gestalt theory at the University of Berlin and even patented a color wheel that was useful in early Gestalt experiments on perception. As a philosopher, he dived deeply into turn-of-the-century epistemology, evaluating, as we discussed in chapter 2, Ernst Mach's psychophysical parallelism in his doctoral dissertation. As discussion in chapter 6 then revealed, Musil's work on Mach was of sufficient quality to attract an offer from the noted intentionalist philosopher Alexius Meinong to do a second doctorate or "habilitation" and so become qualified for a university professorship. Yet the urge towards creative writing was to prove stronger; moreover, Musil did not wish to put off his own investigations of what was happening in the world while he got himself properly qualified for an academic career. In the end, we have Musil's major novel as a report about change in the intellectual firmament, not a career in philosophy and psychology prepared under Meinong's direction in the academy. And we must conclude that Musil chose wisely, for, because of a Jewish wife and a highly developed interest in a broad range of fields, that career was destined to be abbreviated in the German-speaking academy and would certainly not have continued much beyond the early 1930s. As it was, *The Man Without Qualities* was placed on Hitler's list of banned literature in 1938, the year Hitler annexed Austria. In the same year Musil and his wife took themselves across the border to Switzerland, which of course did not participate in hostilities in the Second World War. Musil died there from a stroke in 1942. He was still working on his novel on the day he died.

The ideas put forward in chapter 116 were initially adumbrated in Musil's 1918 essay "Sketch of the Knowledge of a Writer" (see discussion of this essay in chapter 4). The aim of this early essay, broadly speaking, was to give insight into the nature of poetical knowledge, which is to say, to describe the knowledge that literature in its highest and best forms seeks to convey. In carrying out this mission, the essay discusses key aspects of the new thinking that was emerging in Musil's own day and which, from today's vantage point, has points of overlap with the "fourth revolution" in knowledge. The "poetical knowledge" literature discusses is not one or the other of the two aspects of thinking Musil goes on adumbrate in the essay. Rather, Musil's point is that "poetical knowledge" evaluates the tasks confronting thought as a whole. Extending from this essay, we can say that literature—certainly in the form we encounter it under Musil's pen—has utility as a metatheoretical discussion about the nature of thinking at moments of significant sociocultural and intellectual change. In reviewing Musil's novel, we may even say that literature of this type is better positioned to advance such a metatheoretical discussion than other types of formal disclosure which, as with academic discourse (despite the ongoing appellation of academic freedom), are routinely subject to various forms of institutional procedure that constrain and "silo" it, often to its detriment.

As discussion in chapter 4 showed, Musil characterizes the new thinking he observes in his age with the neologisms "ratioïd" and "non-ratioïd." The first term, the ratioïd, is, as Musil conceives it, thinking based on system, rules, and laws, and on repeating instances. The use of the neologism indicates that this thinking is derived from rational logic—and retains the reference to this logic with the base word "ratio"—but is no longer fully subsumable under such rationality. In today's terms, with our own social and cultural development since 1918 in mind, we can say that the ratioïd anticipates the type of codified machine-thinking enshrined in sequences of binary decision-making—"yes" and "no" decisions operationalized in terms of "1" and "0" flow sequences in a unified flow system. The term is thus consistent in its conceptual reach with the digital processes that serve as the basis of the modern computer. Systems theory, where classification begins with the simple difference between outside or *Fremdreferenz* and internal reference or *Eigenreferenz*, is the attempt to apply this kind of binary thought to areas such as the social sciences. There have even been notable attempts to apply systems theory to literature (for example, Schwanitz 1990, Jäger 1994 and, in application to the author Kleist, Theissen 1996). It is fair to say, then, that binary thinking is alive and well in many areas of the humanities today. Musil cannot have known about the machines constructed on the basis of binary logic, even if his intuition as a scientist and engineer told him they were just round the corner. His concern was rather with the cultural effects that this new type of thinking is likely to occasion. And here he provides two insights: the first is that thinking under the guise of the ratioïd will be fundamentally different from non-ratioïd thinking—thinking, as we might now term it, in "analog" mode. This latter type of thinking is not based on repeatable generalized inputs, but rather, as Musil says, on thinking based around the exception, or, as we would now tend to specify, thinking that proceeds from singularity.

The second insight he offers in the essay, as was discussed in chapter 4, is that the ratioïd "domain" (as Musil terms it) will by and large not be able to do ethical work. Indeed, it is a notable feature of the systems theory put forward by Luhmann and others in the social sciences that social theories premised on binary logic entirely lack an ethical component. Ethical work, Musil goes on to say, can only successfully take place in the domain of the non-ratioïd, which is to say, in the mode where it becomes important to estimate the kind of relationship that obtains between the singularity of agency and the multiplicity of the environment with which singular agents "situationally" interact. Note that non-ratioïd does not mean "non-rational." But it does mean that rationality no longer occupies a majority position in guiding determinations in this domain. Rationality, or conceits and procedures derived from reason, will only represent a small part of this non-ratioïd domain.

The important takeaway from Musil's early essay, then, is not just that something like a new era in thought is likely to emerge, but also that, in this new era (let us call it the dawning age of the fourth revolution), we will likely no longer be able to do much, if any, ethical work in the area of thinking that was once inseparably bound up with our rational commitments. Our ethical thinking, then, becomes in an important sense post-rational. Another way of saying the same thing, to put this point in the language of the Frankfurt School, is that key aspects of rational thought in recent modernity have become "instrumentalized." If we do not further isolate—as Musil does—which aspects of thought are the ones open to instrumentalization and which ones are not, we will reach broadly pessimistic conclusions about the prospects for all thinking emerging in the first instance from reason—as Max Horkheimer does in *The Eclipse of Reason* (1944), Theodor Adorno does in *Negative Dialectics* (1966) and both Horkheimer and Adorno do in *The Dialectic of Enlightenment* (1947). This pessimistic strain was enormously influential in European and Anglo-American cultural discussions in the second half of the twentieth century and, as we discussed in chapter 6, has also left its imprint on Musil scholarship. While there can be no doubting the importance of this thought in general terms, the view that Musil ultimately came under its thrall in the last years of his life is mistaken, as I maintain in chapter 6. Musil, I argue, is not finally to be understood as a cultural pessimist. While he well understood the totalizing aspects attaching to instrumental thought, he saw these as emanating overwhelmingly from the domain of the ratioïd. The ratioïd, of course, is the technical, and Musil—as will have become clear from the discussion throughout this volume—had a keen appreciation of its advantages. Yet Musil equally foresaw that we cannot ask the technical area of life to do our ethics for us. This means that the link in intellectual history between reason, *insofar as it enables the appearance of the scientific and the technical*, and ethical thought is no longer valid. Such a link is postulated as central, for example, in Kant's critical philosophy.

The attempt to begin the process of reconceiving how we must now do ethical work—the problem, as Musil calls it in his major novel, of "the right way to live" (*GW* 1:255)—is set out in the vast enterprise called *The Man Without Qualities*. In the language of that novel and specifically in chapter 116 of the novel, Musil characterizes the ethical question his age confronts as the problem of the coexistence of two separate domains for thinking: that which he goes on to term "the figurative," on one side, and "the unequivocal," on the other (*GW* 2:, 593). The figurative—which develops a symbolic language based on metaphor that we associate with the language of art and religion—moves along an axis of particularity. Musil calls this thinking "the gliding logic of the soul" (593). The unequivocal—the rule of lawfulness, exactitude, and system—moves along an axis of exemplarity. Musil locates this second axis of thought at the beginning of humanity: "it arises," he says, "from the exigencies of life where only the single-minded control of circumstances can avert disaster" (593). So, as it turns out, we need both types of thinking; each, in its own domain, has its proper uses and can achieve good results. Problems arise when we can no longer keep both types of thinking in focus and can no longer avail ourselves of the advantages of both. And that such a problem is indeed the case in Musil's time, was the problem situation Musil's major novel discusses. In the course of recent human development, of which the Austrian version of the human is a case in point, the two "trees of life"—the figurative and the unequivocal—have been thrown out of alignment. Projected on to the main character Ulrich, the bearer of this cultural misshapenness, the narrator concludes: "His development had evidently split into two tracks, one running on the surface in daylight, the other in the dark below and closed to traffic, so that the state of moral arrest . . . might simply be the result of [the] failure to bring these two tracks together" (*GW* 2:593).

## Musil and the Thought of Today

In the last part of this concluding chapter I take up a strand of the argument in chapter 116 that is proving fruitful in contemporary intellectual debates. I do not make the argument that Musil is seen as a crucial source of inspiration for this argument. Debates in the human sciences frequently come about not because of a particular finding in a single area of knowledge or on the strength of a single piece of exegesis, though I do not wish to exclude this possibility. Rather it seems to be the case—much as Thomas Kuhn (1996) claims about the advancement of knowledge in the physical sciences—that a problem comes to light at a certain moment in time because many minds in many different fields of knowledge are suddenly exercised by it, can no longer postpone dealing with it, and because a way forward is urgently sought in response to it. Such, I believe, is the case with the task of ethical thinking in post-rational and post-religious communities today. The case I make in relation to the novel *The Man*

*Without Qualities* is that Robert Musil got to this issue in its current complexity before just about everyone else. His unfinished novel is a masterpiece not only because of its qualities as fiction, which are considerable, but also because Musil provided in it a detailed sketch of the big problem of ethics in a post-religious age and brought it to wider attention. I believe we should not underestimate how the best literature can do this important work of interpretative disclosure, and how the most exacting and intelligent literature can distill the essence of a problem long before almost everyone else knows it is there and long before the compartmentalized arrangement of our knowledge factories, including those at our universities, allow it even to be recognized as a problem.

The contemporary debate I allude to is that between exemplarity and singularity. I draw from Michèle Lowrie and Susanne Lüdemann's recent collection of essays under the title *Exemplarity and Singularity* (2015) in conveying a sense of it. This debate is consistent with Musil's presentation of the features of thinking under the heading of "the two trees of life"—the unequivocal strategy in thinking, on one side, and the figurative on the other. Unequivocal thinking is exemplary in that it presents as an attempted integration—it presents a modular truth and a logic of subsumption under this modular truth, be it along horizontal or "flat" semantic lines or vertical and hierarchical ones. The aim of this thinking is to reduce complexity arising from difference and to increase interpretative tolerance for certain broader ways of understanding the appearances. In all cases exemplification serves the purpose of establishing what knowledge is, since, as Lowrie and Lüdermann correctly tell us, knowledge "is defined as the subsumption of particular events or appearances under general laws or rules" (Lowrie/Lüdermann 2015, 4). Immanuel Kant provides a succinct summary of the relationship between exemplarity and singularity in the *Critique of Pure Reason* (also quoted by Lowrie and Lüdermann): "If understanding (*Verstand*) in general is defined as the faculty of rules, judgment (*Urteilskraft*) may be termed the faculty of *subsuming* under these rules; that is, of distinguishing whether something does or does not stand under a given rule (casus datae legis)" (original emphasis).[3] Now, when Musil tells us that the "two trees of life" are out of alignment, he is also saying, with Kant, that we no longer accept the terms of the background relationship governing rules and particulars, which is to say, we do not quite know any longer what constitutes the proper terms of knowledge. This problem of ascertaining lawfulness arises out of the deeper problem of the warrant for second-order lawfulness in

[3] Kant 1990, 2:184: B 172 (emphasis in the original): "Wenn der Verstand überhaupt als das Vermögen der Regeln erklärt wird, so ist Urteilskraft das Vermögen, unter Regeln zu *subsumieren*, d.i. zu unterscheiden, ob etwas unter einer gegebenen Regel (casus datae legis) stehe, oder nicht."

the faculty of judgment. Kant sought to avoid an infinite regress here by claiming that knowledge of lawfulness rested on a "hidden art in the depths of the human soul" (cf. Bowie 1997, 57). But this reference to talent does not get us very far.

The problem of determining the background relationship governing rules and particulars is clearly a crushing problem when a dictator appears on the scene—an event that Musil's novel foreshadows. But even if we look away from the issue of the emergence of dictators, we still have a problem. Nietzsche launched his critique of Kant's categorical imperative on the basis of it—the categorical imperative being a notable attempt to articulate a general moral law while at the same time subsuming particulars in their singularity beneath it. Nietzsche thought that this imperative does not do justice to the particular sense of life felt by individuals under the conditions of modernity, and Musil thoroughly agreed with him. Foucault, too, as Lowrie and Lüdemann point out, saw "the entry of the individual (and no longer the species) into the field of knowledge" as a basic issue. The modern problem we have once we take the entry of the individual into the field of knowledge seriously is how we measure the desire for exemplarity, which is the source of our normative commitments as well as the source from which we derive a workable definition of scientific knowledge, against our desire for singularity, for a basic valorization of our individual lives, our passionate, often idiosyncratic, uncontainable selves. To put the emphasis too much on singularity would appear to risk these normative and scientific commitments. To put the emphasis too squarely on exemplarity—something that we might do in the ratioïd or unequivocal domain—risks, as Musil tells us, a cultural malaise. Yet we are taxed by neither distortion so long as the need to come up with ethical responses to our world is not felt to be pressing. So this, now, is the major point: ethics is the one domain of thought where both exemplarity and singularity have a foundational role to play in its constitution and where both must be kept in focus. If we want a proper ethics, in other words, we must seek to put exemplarity and singularity, respectively and together, onto an even playing field.

The attempt to do this is to my mind what the best imaginative literature seeks to do. For its protagonists are not true individuals or true particulars but, like Ulrich in *The Man Without Qualities*, exemplars, bearers of characteristics that have notional or representative weight or value for us—characteristics whose weight or value, as they are in Musil's novel, might be deepened by the significance of the moments in time they pass through. These exemplars are not the block from which the chip is made, but in the making of the chip there is something beyond the singular cut into wood or stone fashioning their individuality that turns this chip into something we recognize as a common problem, a common virtue, a wider excellence or a deeper failing. Friedrich Schleiermacher, whose ethical thought anticipates some of the discussion here, spoke in his *Lectures on Philosophical Ethics* of "a common particularity" in relation to such exemplars (Schleiermacher 2002, 23). In this mixing of singular instance and example we walk upon the terrain of the best literature. Because, when we read literature or view an artwork, we are led beyond our individual selves without submerging these selves in the morass of generality that would dissolve their particularity and all thought of the gift of individual personality along with it. The capacity of literature or art to move us arises precisely on account of the preservation of the sense of personal life and, at the same time, the connectedness of that life with something greater than it. To make a new point—but one that Musil's novel leads us to—the peregrinations in which the reader participates in imaginative literature in its best forms is strikingly similar to the terrain over which we move when we transact our ethical commitments. For ethics strikes roots in the special mixed area of our thinking where we keep normative exemplarity in dialog with singularity—a thinking where the exceptional instance rubs up against the general abstract laws that govern and condition human forms of association.

Musil's great feat is not only to arrive at a name for this exceptional exemplarity—he assigns his protagonist, as a "man without qualities," to that class of individuals without a class, who, on the strength of their particularity, stand outside the class of themselves. Musil's feat is also to arrive at a description of the type of work this literature seeks to perform. Ulrich, Musil's protagonist, puts forward to his interlocutors in chapter 116 a proposal for a "secretariat of precision and soul." That this proposal falls on deaf ears in the novel is only to be expected: his companions are already too far down the path of one-sided technical thinking to find interest in more salutary ways to respond to their world. Though Musil never finished his great novel, his unpublished notes tell us unmistakably that the results of the cultural development he was observing were that "all lines [would] lead to war" (*GW* 5:1851). But—setting this projected disastrous end to the novel to one side for a moment—in the vision of the type of work that needs to be carried out in order to realize the virtues arising from both exemplarity and singularity—work combining the attributes of both "precision and soul"—we have, notwithstanding that disastrous prospect of war, one of the most profound expressions of the type of thinking we are called upon to cultivate if we are to realize ethical outcomes.

"Precision" and "soul" do not describe the respective methodologies of the ratioïd and non-ratioïd domains so much as their results. When we carry out thinking about exemplarity in the ratioïd domain our goal will be to reach a high level of precision. We now take for granted the computational devices of the fourth revolution that help us attain such precision. In the non-ratioïd domain, by contrast, the goal, as we have seen, is to reach ethical outcomes—the type of intimate, painstakingly hard-won knowledge about ourselves in relation to the world we find around us that we associate with the word "soul." "Truth" would be another word for it. In estimating what results when we ignore the question of the proper relation between the ratioïd and the non-ratioïd domains, Musil did us good service. What follows a cultural malaise of the profound type he was describing, Musil thought, would be war. For the individual who can only go to war with herself one outcome might be suicide. We recall that this is a possibility Musil raised with us at the start of his novel.

The philosopher Karl-Otto Apel took up Musil's warnings about an impending cultural malaise in *The Man Without Qualities* in an important essay that appeared in 1968 (reproduced in Apel 1985). In this essay, writing on the eve of the flowering of postmodernism, Apel took issue with historicist attempts to stabilize the relationship between what Musil had called the two "trees of life." As Apel contended about the novel and its protagonist, Musil had pointed to the issue of "nihilistic 'historicism'"—the cultural issue that arises when we lack a proper way of reconciling the thinking work of these two domains. Apel's description of the work of these domains is notably Musilian: "It appears to me that humans have fundamentally two equally important, not identical, but rather *complementary* cognitive interests: 1) one which is determined by the necessity of a technological praxis on the basis of insight into natural laws; 2) one which is determined by the necessity of a morally relevant social praxis" (Apel 1985, 331). Looking back over the last fifty years, it appears that we have lived through the worst moments of this "nihilistic 'historicism.'" Musil's vision has thus come to pass in two senses: first, we still lack a convincing metatheoretical account of how to unite the virtues of the two areas our thinking depends on (Apel's two points above, which map seamlessly onto Musil's two "trees of life"); second, we now see all too well what can happen in the wider sociopolitical environment when we overlook or otherwise downgrade the importance of the domain area in which our ethical work is transacted. What happens, among other things, is that our leaders also lose touch with ethical thinking. The major issue here appears to relate both to the background terms of knowledge (the background relation between rules/exemplarity and exceptions/singularity in second-order judgment) as well as what flows from this—the absence of something like a tradition that allows rules to be rules long enough to stabilize perspectives and the expectations raised from them. This is the space, as I see it, into which the populism we see all around us today rushes like an advancing tide.

The issues here are too complex to unravel properly here. But at least this much, following Apel, can be said. We need a better "mediation of social-scientific 'explanation' and historical-hermeneutical 'understanding' of meaning-traditions" than we currently have. "Social-scientific 'explanation'" is Apel's name for the method of thinking performed under the ruling notion of system and instrumentalized precision in the ratioïd domain. We typically associate this thinking with the natural and physical sciences. Remember Musil's helpful reassurance that this is one "tree of life" (it grows, sustains itself, and bears fruit). "'Understanding' of meaning-traditions"—the other "tree of life"—is work typically done in the *Geisteswissenschaften*—the humanities and the social sciences of today. Yet this work has recently confronted, and still confronts, a profound challenge. The destabilization of the notion of tradition described under the heading of postmodernism indicates that we no longer have access to the "dogmatic-normative (institutionally established and socially binding) 'application'" of tradition that once allowed us to secure social progress under the banner of Enlightenment (cf. Apel 336). This is the judgment crisis alluded to before. Our judgment is now both post-Enlightenment and, as Musil also foresaw, absorbing insights from Nietzsche, post-rational and post-religious. To answer the judgment crisis, and thereby to re-temper the mediation that must always apply to the work of "explanation" typical of the physical sciences and that of "understanding" typical of the human sciences, is the urgent task we have before us today.

Musil thought of the work confronting us as being the work of a "higher humanism"—"the attempt to fuse together, once they have been carefully distinguished from each other, these two great halves of life" (*GW* 2:593).

# References

## Editions of Musil's Works

Musil, Robert. 1978. *Gesammelte Werke in Neun Bänden.* Edited by Adolf Frisé. Reinbek bei Hamburg: Rowohlt. Cited throughout with the abbreviation *GW*, volume number, and page reference.

———. 1980. *Beitrag zur Beurteilung der Lehren Machs und Studien zur Technik und Psychotechnik* [1908]. Reinbek bei Hamburg: Rowohlt Verlag.

———. 1983. *Tagebücher.* Edited by Adolf Frisé. Reinbek bei Hamburg: Rowohlt. Cited throughout using the abbreviation *TB*, volume number, and page reference.

———. 1992. *Der literarische Nachlass.* CD-Rom. Edited by Friedbert Aspetsberger, Karl Eibl, and Adolf Frisé. Reinbek: Rowohlt Verlag.

## Secondary Literature

Adams, Dale. 2011. *Die Konfrontation von Denken und Wirklichkeit: Die Rolle und Bedeutung der Mathematik bei Robert Musil, Hermann Broch und Friedrich Dürrenmatt.* St. Ingbert: Röhrig.

Adorno, Theodor. 1966. *Negative Dialektik.* Frankfurt am Main: Suhrkamp.

Albertsen, Elisabeth. 1968. *Ratio und "Mystik" im Werk Robert Musils.* Munich: Nymphenburger.

Anderson, T. W. 1958. *An Introduction to Multivariate Statistical Analysis.* New York: Wiley.

Apel, Karl-Otto. 1985. "Perspectives for a General Hermeneutic Theory." *The Hermeneutics Reader: Texts of the German Tradition from the Enlightenment to the Present.* Edited by Kurt Mueller-Vollmer, 320–45. London: Continuum.

Arntzen, Helmut. 1980. *Musil-Kommentar sämtlicher zu Lebzeiten erschienener Schriften außer dem Roman "Der Mann ohne Eigenschaften."* Munich: Winkler Verlag.

———. 1982. *Musil-Kommentar zum Roman "Der Mann ohne Eigenschaften."* Munich: Winkler Verlag.

Arvon, Henri. 1970. "Robert Musil und der Positivismus." In *Robert Musil: Studien zu seinem Werk.* Edited by Karl Dinklage, Elisabeth Albertsen, and Karl Corino, 200–213. Reinbek bei Hamburg: Rowohlt.

Bahr, Hermann. 1981. "Das unrettbare Ich." In *Die Wiener Moderne: Literatur, Kunst und Musik zwischen 1890 und 1910.* Edited by Gotthart Wunberg, 147–48. Stuttgart: Reclam.

Balázs, Béla. 1924. *Der sichtbare Mensch.* Vienna: Deutsch-österreichischer.

Barad, Karen. 2007. *Meeting the Universe Halfway: Quantum Physics and the Entanglement of Matter and Meaning.* Durham, NC: Duke University Press.

Berghahn, Wilfried. 1963. *Robert Musil.* Reinbek bei Hamburg: Rowohlt.

Bergson, Henri. 1992. *The Creative Mind: An Introduction to Metaphysics.* New York: Citadel Press.

Bhatti, Anil, Dorothee Kimmich, Albrecht Koschorke, Rodolf Schlögl, and Jürgen Wertheimer. 2011. "Ähnlichkeit: Ein kulturtheoretisches Paradigma." *Internationales Archiv für Sozialgeschichte der deutschen Literatur* 36, no. 1: 233–47.

Böhme, Hartmut. 1974. *Anomie und Entfremdung: Literatursoziologische Untersuchungen zu den Essays Robert Musils und seinem Roman "Der Mann ohne Eigenschaften."* Kronberg i. Ts.: Scriptor.

———. 1990. "'Erinnerungszeichen an unverständliche Gefühle.'" In *Robert Musil: Vereinigungen.* 185–221. Frankfurt am Main: Suhrkamp.

Bowie, Andrew. 1997. *From Romanticism to Critical Theory: The Philosophy of German Literary Theory.* London: Routledge.

Brynjolfsson, B., and A. McAfee. 2016. *The Second Machine Age.* New York: Norton.

Bunge, Mario. 1979. *Causality and Modern Science.* New York: Dover.

Calinescu, Matei. 1987. *Five Faces of Modernity.* Durham, NC: Duke University Press.

Cassirer, Ernst. 1969. *Substanzbegriff und Funktionsbegriff: Untersuchungen über die Grundfragen der Erkenntniskritik* [1910]. Darmstadt: Wissenschaftliche Buchgesellschaft.

Coetzee, J. M. 1992. *Doubling the Point: Essays and Interviews.* Edited by David Attwell. Cambridge, MA: Harvard University Press.

Corino, Karl. 2005. *Robert Musil: Eine Biographie.* Reinbek: Rowohlt Verlag.

Dahan-Gaida, Laurence. 1993/4. "Die Wärmetheorie bei Robert Musil." *Musil-Forum* 19/20: 117–31.

Dressler-Brumme, Charlotte. 1987. *Nietzsches Philosophie in Musils Roman "Mann ohne Eigenschaften."* Frankfurt am Main: Athenäum.

Emter, Elisabeth. 1985. *Literatur und Quantentheorie: Die Rezeption der modernen Physik in Schriften zur Literatur und Philosophie deutschsprachiger Autoren (1925–1970).* Bern: Walter de Gruyter.

Feyerabend, Paul. 1995. "Erkenntnis und Praxis: Gefahren der Abstraktion, Fallgruben des Realen." *Lettre international* 29 (2): 20–23.

Floridi, Luciano. 2014. *The Fourth Revolution: How the Infosphere Is Reshaping Human Reality.* Oxford: Oxford University Press.

Frank, Manfred. 1988. "Remythisierte Erkenntniskritik (Robert Musil)." In M.F., *Gott im Exil: Vorlesungen über die neue Mythologie*, 2:315–32. Frankfurt am Main: Suhrkamp.

Freed, Mark. 2011. *Robert Musil and the NonModern.* New York: Continuum.

Fricke, Gerhard. 1929. *Gefühl und Schicksal bei Heinrich von Kleist.* Berlin: Juncker and Dünnhaupt.

Gadamer, Hans-Georg. 2010. *Wahrheit und Methode: Grundzüge einer philosophischen Hermeneutik*. Tübingen: Mohr Siebeck.

Gies, Annette. 2003. *Musils Konzeption des "Sentimentalen Denkens": "Der Mann ohne Eigenschaften" als literarische Erkenntnistheorie*. Würzburg: Königshausen & Neumann.

Gittell, Ross, and Avis Vidal. 1988. *Community Organizing: Building Social Capital as a Developmental Strategy*. Thousand Oaks, CA: Sage Publications.

Glander, Kordula. 2005. "Licht und Farbe in Texten Robert Musils." In *Alle Welt ist medial geworden. Literatur, Technik, Naturwissenschaft in der klassischen Moderne*, edited by Matthias Luserke-Jaqui, 127–36. Tübingen: Narr Francke.

Gnam, Andrea. 1998. *Die Bewältigung der Geschwindigkeit: Robert Musils Roman "Der Mann ohne Eigenschaften" und Walter Benjamins Spätwerk*. Munich: Wilhelm Fink.

Goldgar, Harry. 1969. "The Square Root of Minus One: Freud and Robert Musil's *Törleß*." *Comparative Literature* 17, no. 2: 117–32.

Graff, Gerald. 1977. "Fear and Trembling at Yale." *American Scholar* 46/4: 467–78.

Guillory, John. 1993. *Cultural Capital: The Problem of Literary Canon Formation*. Chicago, IL: University of Chicago Press.

Habermas, Jürgen. 1981. *Die Moderne—ein unvollendetes Projekt: Philosophisch-politische Aufsätze*. Frankfurt am Main: Suhrkamp.

Hajduk, Stefan. 2000. *Die Figur des Erhabenen: Robert Musils ästhetische Transgression der Moderne*. Würzburg: Königshausen & Neumann.

Heidegger, Martin. 1962. *Kant and the Problem of Metaphysics*. Translated by James S. Churchill. Bloomington: Indiana University Press.

———. 1995. *The Fundamental Concepts of Metaphysics: World, Finitude, Solitude*. Translated by William McNeill and Nicholas Walker. Bloomington: Indiana University Press.

———. 2002. *Off the Beaten Track*. Edited and translated by Julian Young and Kenneth Haynes. Cambridge: Cambridge University Press.

Heisenberg, Werner. 2006. "Was bedeutet Verstehen in der theoretischen Physik? Kant Demokrit Plato: Die Physik der Elementarteilchen." In *Die Verknüpfung von Physik und Philosophie: Originaltonaufnahmen 1951–1967*. Audiobook. Berlin: Supposé.

Herrmann, Theo. 1982. "Ganzheitspsychologie und Gestalttheorie." In *Geschichte der Psychologie*, edited by Heinrich Balmer, 573–658. Vol. 1. Weinheim, Basel: Beltz.

Hoffmeister, Werner. 1965. "Die erlebte Rede im Werk Robert Musils." In W.H.: *Studien zur erlebten Rede bei Thomas Mann und Robert Musil*, 86–159. The Hague: Mouton.

Hoheisel, Claus. 2009. *Das Doppelgesicht der Natur: Naturwissenschaftliche Aspekte des "anderen Zustands" in Robert Musils Roman "Der Mann ohne Eigenschaften."* Berlin: Europäischer Universitätsverlag.

Honold, Alexander. 1993. "Die verwahrte und die entsprungene Zeit: Paul Kellers *Ferien vom Ich* und die Zeitdarstellung im Werk Robert Musils." *Deutsche Vierteljahrsschrift* 67/2: 302–21.

Horkheimer, Max. 1947. *The Eclipse of Reason.* Oxford: Oxford University Press.

Horkheimer, Max, and Adorno, Theodor. 1944. *Dialektik der Aufklärung: Philosophische Fragmente.* New York: Social Studies Association.

Hüppauf, Bernd-Rüdiger. 1971. *Von sozialer Utopie zur Mystik: Zu Robert Musils "Der Mann ohne Eigenschaften."* Munich: Wilhelm Fink.

Jäger, Georg. 1994. "Systemtheorie und Literatur: Teil I, Der Systembegriff der empirischen Literaturwissenschaft." *Internationales Archiv für Sozialgeschichte der deutschen Literatur* 19/1: 95–125.

Jameson, Fredric. 2002. *A Singular Modernity: Essay on the Ontology of the Present.* London: Verso.

Jappe, Lilith. 2011. *Selbstkonstitution bei Robert Musil und in der Psychoanalyse: Identität und Wirklichkeit im "Mann ohne Eigenschaften."* Munich: Wilhelm Fink.

Jäßl, Gerolf. 1963. *Mathematik und Mystik in Robert Musils Roman "Der Mann ohne Eigenschaften."* Diss. Munich University.

Jonsson, Stefan. 2000. *Subject without Nation: Robert Musil and the History of Modern Identity.* Durham, NC: Duke University Press.

Kant, Immanuel. 1990. *Werke in sechs Bänden.* Edited by Wilhelm Weischedel. Darmstadt: Wissenschaftliche Buchgesellschaft.

Kassung, Christian. 2001 *EntropieGeschichten: Robert Musils "Der Mann ohne Eigenschaften" im Diskurs der modernen Physik.* Munich: Wilhelm Fink.

Kittler, Friedrich A. 1990. *Aufschreibesysteme 1800/1900* (1985). Published in English as *Discourse Networks 1800/1900.* Translated by Michael Metteer, with Chris Cullens. Stanford, CA: Stanford University Press.

Kochs, Angela. 1996. *Chaos und Individuum: Robert Musils philosophischer Roman als Vision der Moderne.* Munich: Karl Alber Freiburg.

Kollmann, Franz Gustav. 2007. *Robert Musil und die Mathematik.* Stuttgart: Steiner.

Könneker, Carsten. 2001. *"Auflösung der Natur. Auflösung der Geschichte": Moderner Roman und NS-"Weltanschauung" im Zeichen der theoretischen Physik.* Stuttgart: Metzler.

Kramnick, Isaac, ed. 1995. "Introduction." In *The Portable Enlightenment Reader.* London: Penguin, ix–xiii.

Krommer, Alex, and Albert Kümmel. 1993/4. "Ordnung verlangt nach Zerrissenwerden: Skizze einer informationstheoretischen Deutung des *Mann ohne Eigenschaften.*" *Musil-Forum* 19/20: 158–64.

Kuhn, Reinhard. 1976. *The Demon of Noontide: Ennui in Western Literature.* Princeton, NJ: Princeton University Press.

Kuhn, Thomas S. 1996. *The Structure of Scientific Revolutions.* Chicago, IL: University of Chicago Press.

Kümmel, Albert. 2001. *Das MoE-Programm: Eine Studie über geistige Organisation.* Munich: Wilhelm Fink.

Laermann, Klaus. 1970. *Eigenschaftslosigkeit: Reflexionen zu Musils Roman "Der Mann ohne Eigenschaften."* Stuttgart: Metzler.

Landgraf, Edgar, Gabriel Trop, and Leif Weatherby, eds. 2019. *Posthumanism in the Age of Humanism: Mind, Matter, and the Life Sciences after Kant.* New York: Bloomsbury Academic.

Lefebvre, Henri. 1991. *Production de l'espace* (1974). In English: *The Production of Space.* Translated by Donald Nicholson-Smith. Oxford, UK: Blackwell.

Lessing, Gotthold Ephraim. 1766. *Laokoön.* Berlin: C. F. Voss.

Lin, Nan, Karen Cook, and Ronald S. Burt, eds. 2001. *Social Capital: Theory and Research.* New York: Aldine de Gruyter.

Lowrie, Michèle, and Susanne Lüdemann, eds. 2015. *Exemplarity and Singularity: Thinking through Particulars in Philosophy, Literature, and Law.* London: Routledge.

Luft, David. 1980. *Musil and the Crisis of European Culture 1880–1942.* Berkeley: University of California Press.

Luhmann, Niklas. 1997. *Die Gesellschaft der Gesellschaft.* Frankfurt am Main: Suhrkamp.

———. 1998. *Observations on Modernity.* Translated by William Whobrey. Stanford, CA: Stanford University Press.

———. 2000. *Vertrauen: Ein Mechanismus der Reduktion sozialer Komplexität.* Stuttgart: Lucius & Lucius.

Luserke, Matthias. 1987. *Wirklichkeit und Möglichkeit: Modaltheoretische Untersuchung zum Werk Robert Musils.* Frankfurt am Main: Peter Lang.

Mach, Ernst. 1985. *Die Analyse der Empfindungen und das Verhältnis des Physischen zum Psychischen.* Darmstadt: Wissenschaftliche Buchgesellschaft.

———. 1991. *Erkenntnis und Irrtum: Skizzen zur Psychologie der Forschung.* Darmstadt: Wissenschaftliche Buchgesellschaft.

MacIntyre, Alastair. 1996. *A Short History of Ethics: A History of Moral Philosophy from the Homeric Age to the Twentieth Century.* London: Routledge & Kegan Paul.

Mandelkow, Valentin. 1999. *Der Prozeß um den "ennui" in der französischen Literatur und Literaturkritik.* Würzburg: Königshausen & Neumann.

McBride, Patrizia. 2006. *The Void of Ethics: Robert Musil and the Experience of Modernity.* Evanston, IL: Northwestern University Press.

Mehigan, Tim. 1995. "Moral und Verbrechen: Über Musils intellektuelle Position." *Wirkendes Wort* 2:227–40.

———. 1997. "Robert Musil, Ernst Mach und das Problem der Kausalität." *Deutsche Vierteljahrsschrift für Literaturwissenschaft und Geistesgeschichte* 71:264–87.

———. 2001. *Robert Musil.* Stuttgart: Reclam.

———. 2005. "Musil mit Luhmann: Das Problem des Vertrauens in Musils *Mann ohne Eigenschaften.*" In *Musil anders: Neue Erkundungen eines Autors zwischen den Diskursen.* Edited by Gunther Martens, Clemens Ruthner, and Jaak De Vos, 45–59. Bern: Peter Lang.

———. 2007. "Postcritical Criticism: Musil as an Example." In *Literarische Wertung und Kanonbildung*, edited by Nicholas Saul and Ricarda Schmidt, 50–65. Berlin: de Gruyter/Königshausen & Neumann.

———. 2010. "Robert Musil." *Praktizierte Intermedialität: Deutsch-französische Porträts von Schiller bis Goscinny/Uderzo.* Edited by Fernand Hörner, Harald Neumeyer, and Bernd Stiegler, 87–103. Bielefeld: Transcript-Verlag.

———. 2013. "Robert Musil: *Der Mann ohne Eigenschaften* (1930; 1933)." In *Literatur und Wissen: Ein interdisziplinäres Handbuch.* Edited by Roland Borgards, Harald Neumeyer, Nicolas Pethes, and Yvonne Wübben, 395–400. Stuttgart: Metzler.

———. 2014. "Nachwort: 'Eine Episode von mehr als persönlicher Wichtigkeit.'" In *Robert Musil: Vereinigungen.* Edited by Tim Mehigan, 121–43. Stuttgart: Reclam.

———. 2017. "Robert Musils Essayistik." *Essayistik der Moderne (1918–1950).* Edited by Michael Ansel, Jürgen Egyptien, and Hans-Edwin Friedrich, 46–59. Amsterdam: Rodopi.

———. 2018. "Zum multivariaten Korrespondenzprojekt Robert Musils. Am Beispiel von *Der Mann ohne Eigenschaften* and *Die Amsel.*" *Neue Beiträge zur Germanistik* 17/1: 105–23.

Müller, Gerd. 1971. *Dichtung und Wissenschaft: Studien zu Robert Musils Romanen "Die Verwirrungen des Zöglings Törleß" und "Der Mann ohne Eigenschaften."* Uppsala: Acta Universitatis Upsaliensia.

Müller-Seidel, Walter. 1961. *Versehen und Erkennen.* Cologne: Böhlau.

Nagel, Thomas. 1986. *The View from Nowhere.* New York: Oxford University Press.

Nietzsche, Friedrich. 1968. *Nietzsche Werke: Kritische Gesamtausgabe.* Edited by Giorgio Colli and Mazzino Montinari. Berlin: Walter de Gruyter.

———. 1992. *Der Wille zur Macht.* Frankfurt am Main: Insel-Verlag.

Nübel, Birgit. 2006. *Robert Musil: Essayismus als Selbstreflexion der Moderne.* Berlin: Walter de Gruyter.

Panofsky, Erwin. 1991. *Perspective as Symbolic Form.* New York: Zone Books.

Pelmter, Andrea. 2008. *"Experimentierfeld des Seinkönnens"—Dichtung als "Versuchsstätte": Zur Rolle des Experiments im Werk Robert Musils.* Würzburg: Königshausen & Neumann.

Pieper, Hans-Joachim. 2002. *Musils Philosophie: Essayismus und Dichtung im Spannungsfeld der Theorien Nietzsches und Machs.* Würzburg: Königshausen & Neumann.

Pinker, Steven. 2015. *The Language Instinct: How the Mind Creates Language.* London: Penguin.

Pott, Hans-Georg. 2013. *Kontingenz und Gefühl: Studien zu/mit Robert Musil.* Munich: Wilhelm Fink.

Reinhardt, Stephan. 1969. *Studien zur Antinomie von Intellekt und Gefühl in Musils Roman "Der Mann ohne Eigenschaften."* Bonn: Bouvier.

Renner, Rolf. 2007. "Intermediality and the Simulation of Space." *Seminar: A Journal of Germanic Studies* 43/4: 385–97.

Roth, Marie-Louise. 1972. *Robert Musil: Ethik und Ästhetik.* Munich: List.

Schleiermacher, Friedrich. 2002. *Lectures on Philosophical Ethics.* Edited by Robert B. Louden. Translated by Louise Adey Huish. Cambridge: Cambridge University Press.

Schmidt, Jochen. 1975. *Ohne Eigenschaften: Eine Erläuterung zu Musils Grundbegriff.* Tübingen: Niemeyer.

Scholz, Frithard. 1982. *Freiheit als Indifferenz: Alteuropäische Probleme mit der Systemtheorie Niklas Luhmanns.* Frankfurt am Main: Suhrkamp.

Schöne, Albrecht. 1961. "Zum Gebrauch des Konjunktivs bei Robert Musil." *Euphorion* 55:196–220.

Schraml, Wolfgang. 1994. *Relativismus und Anthropologie: Studien zum Werk Robert Musils und zur Literatur der zwanziger Jahre.* Munich: Eberhard.

Schwab, Klaus. 2017. *The Fourth Industrial Revolution.* London: Portfolio Penguin.

Schwanitz, Dietrich. 1990. *Systemtheorie und Literatur: Ein neues Paradigma.* Wiesbaden: Springer.

Scruton, Roger. 1994. *Modern Philosophy: An Introduction and Survey.* London: Allen Lane Press (Penguin).

Sebastian, Thomas. 2005. *The Intersection of Science and Literature in Musil's "The Man Without Qualities."* Rochester, NY: Camden House.

Specht, Rainer. 1990. *Rationalismus: Geschichte der Philosophie in Text und Darstellung.* Edited by Rüdiger Bubner. Vol. 5. Stuttgart: Reclam.

Theissen, Bianca. 1996. *Bogenschluss: Kleists Formalisierung des Lesens.* Freiburg i. Br.: Rombach.

Toulmin, Stephen. 2001. *Return to Reason.* Cambridge, MA: Harvard University Press.

Wagner-Egelhaaf, Martina. 1989. *Mystik der Moderne: Die visionäre Ästhetik der deutschen Literatur im 20. Jahrhundert.* Stuttgart: Metzler.

Wellmer, Albrecht. 1990. "Metaphysics and the Moment of Its Fall." In *Literary Theory Today*, edited by Peter Collier and Helga Geyer-Ryan, 35–49. Ithaca, NY: Cornell University Press.

Wellmon, Chad. 2015. *Organizing Enlightenment: Information Overload and the Invention of the Modern Research University.* Baltimore, MD: Johns Hopkins University Press.

Wilkerson, T. E. 1976. *Kant's Critique of Pure Reason: A Commentary for Students.* London: Oxford University Press.

Willemsen, Roger. 1985. *Robert Musil: Vom intellektuellen Eros.* Munich: Piper.

Wolf, Norbert. 2011. *Kakanien als Gesellschaftskonstruktion: Robert Musils Sozioanalyse des 20. Jahrhunderts.* Vienna: Böhlau.

Wood, James. 2008. *How Fiction Works.* London: Jonathan Cape.

# Index